Library of Congress Cataloging-in-Publication Data

Rickettsial infection and immunity / edited by Burt Anderson, Herman
    Friedman, and Mauro Bendinelli.
        p.   cm. -- (Infectious agents and pathogenesis)
    Includes bibliographical references and index.
    ISBN 0-306-45528-5
    1. Rickettsial diseases--Immunological aspects.   I. Anderson,
Burt.   II. Friedman, Herman, 1931-   . III. Bendinelli, Mauro.
IV. Series.
    [DNLM: 1. Rickettsiaceae Infections--immunology.   WC 600 R5388
1997]
QR201.R59R53   1997
616.9'22079--dc21
DNLM/DLC
for Library of Congress   97-22825
                                                              CIP

QR
201
.R59
R53
1997

ISBN 0-306-45528-5

© 1997 Plenum Press, New York
A Division of Plenum Publishing Corporation
233 Spring Street, New York, N.Y. 10013

http://www.plenum.com

10 9 8 7 6 5 4 3 2 1

Printed in the United States of America

Contributors

BURT ANDERSON • Department of Medical Microbiology and Immunology, College of Medicine, University of South Florida, Tampa, Florida 33612

OSWALD G. BACA • Department of Biology, The University of New Mexico, Albuquerque, New Mexico 87131

PHILIPPE BROUQUI • Unité des Rickettsies, CNRS-UPRES-A, Faculté de Médecine, 13385 Marseille cedex 5, France

THOMAS D. CONLEY • Cardiology Section, The University of Oklahoma Health Sciences Center, Oklahoma City, Oklahoma 73104. *Present address:* University of Arkansas for Medical Science, Little Rock, Arkansas 72204

J. STEPHEN DUMLER • Division of Medical Microbiology, Department of Pathology, The Johns Hopkins Medical Institutions, Baltimore, Maryland 21287-7093

CAYLE C. GRAUMANN • Department of Molecular Microbiology and Immunology, School of Medicine, University of Missouri, Columbia, Missouri 65212

KAREN K. HAMILTON • Cardiology Section, The University of Oklahoma Health Sciences Center, Oklahoma City, Oklahoma 73104. *Present address:* Cardiology Associates of Corpus Christi, Corpus Christi, Texas 78412

ROBERT A. HEINZEN • Department of Molecular Biology, University of Wyoming, Laramie, Wyoming 82071-3944

THOMAS R. JERRELLS • Department of Pharmaceutical Sciences, College of Pharmacy, Washington State University, Pullman, Washington 99164-6510

GREGORY A. MCDONALD • Department of Molecular Microbiology and Immunology, School of Medicine, University of Missouri, Columbia, Missouri 65212

LOUIS P. MALLAVIA • Department of Microbiology, Washington State University, Pullman, Washington 99164

MICHAEL F. MINNICK • Division of Biological Sciences, The University of Montana, Missoula, Montana 59812-1002

DIDIER RAOULT • Unité des rickettsies, Centre National de la Recherche Scientifique EP J 0054, Faculté de Médecine, 13385 Marseilles, France

VERONIQUE ROUX • Unité des rickettsies, Centre National de la Recherche Scientifique EP J 0054, Faculté de Médecine, 13385 Marseilles, France

DAVID J. SILVERMAN • Department of Microbiology and Immunology, University of Maryland School of Medicine, Baltimore, Maryland 21201

LEONARD N. SLATER • Infectious Diseases Section, The University of Oklahoma Health Sciences Center, and The Department of Veterans Affairs Medical Center, Oklahoma City, Oklahoma 73104

JOSEPH J. TEMENAK • Department of Molecular Microbiology and Immunology, School of Medicine, University of Missouri, Columbia, Missouri 65212

JENIFER TURCO • Department of Biology, Valdosta State University, Valdosta, Georgia 31698-0015

MATTHEW F. WACK • Infectious Diseases Section, The University of Oklahoma Health Sciences Center, and The Department of Veterans Affairs Medical Center, Oklahoma City, Oklahoma 73104. *Present address:* Infectious Disease of Indiana, Indianapolis, Indiana 46202.

DAVID F. WELCH • Pediatric Infectious Diseases Section, The University of Oklahoma Health Sciences Center, Oklahoma City, Oklahoma 73104. *Present address:* Laboratory Corporation of America, Dallas, Texas 75230

HERBERT H. WINKLER • Laboratory of Molecular Biology, Department of Microbiology and Immunology, University of South Alabama College of Medicine, Mobile, Alabama 36688

Preface

Historically, rickettsial diseases have had a profound impact on civilized man. Epidemic typhus has been one of the most devastating infectious diseases affecting humanity. The effects of this disease were most evident during times of war and famine and influenced the outcome and mortality of many wars fought in Europe and Russia before World War II. During and immediately following World War I, over 3 million lives were lost to epidemic typhus in Europe and present-day Russia. In modern times, insecticides have greatly reduced the spread of epidemic typhus by disrupting transmission from the vector, the human body louse, to man. Nevertheless, the disease is still present in Africa and South America. The etiologic agent of epidemic typhus, *Rickettsia prowazekii*, has been isolated from the flying squirrels, *Glaucomys volans*, and their ectoparasites in the southeastern United States. Thus, a reservoir of *R. prowazekii* exists in the United States for transmission to man. Although this naturally occurring reservoir has not resulted in outbreaks of typhus in this region of the country, the possibility must not be discounted. Should conditions of social upheaval or famine occur, epidemic typhus could resurface in the United States, much like the recent resurgence of tuberculosis.

Although the past has provided dramatic historical evidence for the mortality associated with rickettsial diseases, perhaps the most exciting aspect of rickettsial research is currently under way. The advent of modern molecular biology methods has greatly facilitated the study of these obligate intracellular bacteria. Molecular cloning has provided the means by which individual rickettsial genes and the corresponding proteins can be studied without the need of cultivating rickettsiae and purifying them from host cell debris. Polymerase chain reaction amplification has allowed specific genes to be amplified and sequenced without the need to cultivate the organisms. Individual antigens of the etiologic agent of Rocky Mountain spotted fever (*Rickettsia rickettsii*) have been cloned and expressed in bacterial and other expression systems. The resulting recombinant expressed antigens have been used to study the immune response elicited during infection with *R. rickettsii*. In this fashion potential vaccine candidate proteins have been identified and shown to confer protection in animal models. Likewise, molecular tools have advanced recent research efforts aimed at understanding the mechanisms of pathogenesis and the biochemistry of the rickettsiae. The role of individual proteins such as phospholipases, adhesins, invasins, and factors that may affect host-cell signal transduction are areas of current active research.

Molecular techniques have had a great impact on the identification and characterization of new rickettsial pathogens as well. Rickettsiae and rickettsiae-like organisms contain the largest number of recently identified human pathogens of any bacteria. At least six new human pathogens (the agent of human granulocytic ehrlichiosis, the ELB agent, *Ehrlichia chaffeensis*, *Rickettsia japonica*, *Bartonella henselae*, and *B. elizabethae*) have been described so far this decade. Two of these agents, the agent of human granulocytic ehrlichiosis and *E. chaffeensis*, were previously viewed as agents primarily infecting animals. In the inaugural issue of "Emerging Infectious Diseases," the Centers for Disease Control and Prevention names two agents included in this volume (*E. chaffeensis* and *B. henselae*) as major etiologic agents of infectious diseases identified in recent years. The use of polymerase chain reaction amplification of the small subunit ribosomal RNA coupled with sequencing has proven invaluable as a tool to detect and identify new rickettsial pathogens without the need for isolation of the organism. Because rickettsiae are difficult to isolate and cultivate (some have still not been cultivated in the laboratory), the possibility of identifying additional rickettsial pathogens is very real. A number of other new organisms, belonging to almost every genus of rickettsiae or rickettsiae-like organisms, have been isolated or detected in a variety of animal or insect hosts. The ability of these newly described agents to cause disease in man has not yet been extensively studied.

The strict definition of rickettsiae is that they are obligate intracellular bacteria. Until recently the order *Rickettsiales* contained members that could be cultivated in the laboratory on cell-free media. A recently published proposal has been to combine the genus *Rochalimaea* with *Bartonella* and to remove these facultatively intracellular or epicellular bacteria to another order outside the *Rickettsiales*. Regardless, for the purpose of this volume, rickettsiae-like organisms such as *Bartonella* have been included. Perhaps the most exciting recent discovery is the identification of the agent responsible for cat-scratch disease. *Bartonella henselae* (formerly *Rochalimaea henselae*) is most closely related to the agent of trench fever, *B. quintana* (formerly *Rickettsia quintana*). The identification of the etiologic agent of cat-scratch disease was the source of controversy for over 40 years. Compelling evidence provided by PCR, serology, and culture indicates that *B. henselae* is the primary agent responsible for cat-scratch disease. *B. henselae* and *B. quintana* are also opportunistic pathogens of immunocompromised patients and both are etiologic agents of bacillary angiomatosis. Bacillary angiomatosis is a disease seen primarily in AIDS patients and results in unique skin lesions that are characterized by proliferation of vascular endothelial cells. This pseudoneoplasia is also the hallmark of the chronic stage (termed verruga peruana) of infection with *B. bacilliformis* among immunocompetent patients. This novel mechanisms of pathogenesis shared by these organisms is fascinating. The factor or factors from these organisms that mediates angiogenesis remains unidentified but is an area of active research in several prominent laboratories. The possible exchange of DNA coding for virulence factors among these related organisms is under investigation.

Considering the recent interest in rickettsiae and rickettsiae-like organisms, as well as the description of several new pathogens, the editors and authors of this volume feel it is both timely and needed. We anticipate that this volume will be a medium for the cohesive presentation of a variety of different aspects of rickettsial pathogenesis and immunity. Few organisms, if any, offer such fertile grounds for the study of diverse mechanisms of immunity and pathogenesis. It is our hope that this volume summarizes our current knowledge and piques the interest of readers who want to know more about these unusual organisms.

Contents

5. Surface Components of the Spotted Fever Group Rickettsiae

GREGORY A. MCDONALD, CAYLE C. GRAUMANN,
and JOSEPH J. TEMENAK

6. Oxidative Cell Injury and Spotted Fever Group Rickettsiae

DAVID J. SILVERMAN

7. Intracellular Development of *Coxiella burnetii*

ROBERT A. HEINZEN

8. The Identification of Virulence Factors of *Coxiella burnetii*

OSWALD G. BACA and LOUIS P. MALLAVIA

9. Human Granulocytic Ehrlichiosis

    J. STEPHEN DUMLER and PHILIPPE BROUQUI

10. The Immune Response to *Ehrlichia chaffeensis*

    PHILIPPE BROUQUI and J. STEPHEN DUMLER

13. Stimulation of Angiogenesis and Protection from Oxidant Damage: Two Potential Mechanisms Involved in Pathogenesis by *Bartonella henselae* and Other *Bartonella* Species

THOMAS D. CONLEY, MATTHEW F. WACK,
KAREN K. HAMILTON, and LEONARD N. SLATER

Molecular Rickettsiology
A Short History

BURT ANDERSON

1. INTRODUCTION

As recently as 10 years ago, only a handful of papers had been published describing the molecular cloning of genes from any of the rickettsiae. The first nucleotide sequence of a rickettsial gene was reported in 1987. Since that time molecular genetic technology has been incorporated into almost every aspect of rickettsiology including diagnostics, epidemiology, phylogenetic studies, taxonomy, and a variety of studies pertaining to natural history of rickettsial diseases, gene structure and function, and pathogenesis. Molecular analysis, particularly polymerase chain reaction amplification of rickettsial genomes, has resulted in the identification of new rickettsial pathogens. Likewise, the detection and identification of old rickettsial pathogens in new places and in patients with diverse clinical presentations has fueled a resurgence in research interest in these agents that has not been seen in the antibiotic era. The identification of some of these agents as capable of causing "emerging infections" has also served to increase awareness in clinicians. Hence, the number of cases recognized and the diversity of disease attributed to the rickettsiae are likely to increase as diagnostic tests for these agents are simplified and become commercially available.

Although the application of molecular biology techniques to the study of rickettsiology may have lagged behind other bacteria, the impact of such studies has probably been greater. The need to cultivate rickettsiae in embryonating eggs

BURT ANDERSON • Department of Medical Microbiology and Immunology, College of Medicine, University of South Florida, Tampa, Florida 33612.

Rickettsial Infection and Immunity, edited by Anderson *et al.* Plenum Press, New York, 1997.

or tissue culture systems and purify these agents from host cellular debris and organelles makes many of the biochemical and genetic studies much more cumbersome than for other less-fastidious bacteria. By isolating genes encoding antigens, surface proteins, proteins involved in pathogenesis, and proteins encoding key enzymatic functions, it is possible to avoid the time-consuming and laborious tasks of cultivation and purification. Although molecular cloning certainly does not preclude and eliminate the need to confirm gene function in the rickettsiae themselves, the absence of genetic systems still remains as an obstacle to apply classical bacterial genetics to these organisms. Accordingly, molecular techniques such as the use of polymerase chain reaction to amplify individual rickettsial genes has helped in overcoming this obstacle.

In this chapter I have attempted to highlight the application of molecular techniques to rickettsiology primarily in three areas: (1) gene structure and expression, (2) identification of new species, and (3) epidemiology and disease natural history. Although rickettsiology has benefited from the application of molecular techniques to other areas, the three areas on which I focus offer several clear examples that have had a major impact in the field of rickettsiology.

2. GENE STRUCTURE AND EXPRESSION

The first report of the cloning and expression of a rickettsial gene in *E. coli* was by Wood and colleagues in 1983.[1] In that report a cloned segment of *Rickettsia prowazekii* DNA was found to complement a citrate synthase-deficient mutant of *E. coli*. Analysis of the citrate synthase activity, as expressed in *E. coli*, indicated that the enzyme retains the regulatory control mechanism characteristic of the rickettsial enzyme. This report provided the first evidence that at least some rickettsial genes possessed control elements and codon usage that allow expression in *E. coli*. Two years later the same group of researchers reported the cloning of the gene involved in translocation of ATP into *R. prowazekii*.[2] A single cosmid clone, out of 2700 clones screened, accumulated radioactive ATP. This activity was shown to be mediated by the recombinant plasmid harboring 9 kb of *R. prowazekii* DNA. Again, a functionally active rickettsial gene was cloned and expressed in *E. coli*.

Efforts to clone genes encoding rickettsial antigens began to pay off when the cloning and expression of antigenic proteins from *R. tsutsugamushi*[3] and *R. rickettsii*[4,5] were described in 1987. Nucleotide sequence analysis of these cloned genes began to shed light on the reason for the successful expression of rickettsial genes in *E. coli*. The first such report described the sequence of the 17-kDa antigen gene from *R. rickettsii*.[4] The 17-kDa protein is conserved among the typhus- and spotted fever-group rickettsiae.[6] This gene was shown to code for a 477-nucleotide open reading frame capable of coding for a protein of 16,840 Da. The coding region for this antigen was shown to be (A + T) rich as is the whole genome of most rickettsiae and indicating a similar codon usage bias as other (A + T)-rich bacteria.[7] The 17-kDa antigen gene has been sequenced from most species of spotted fever- and typhus-group rickettsiae and has been used extensively for identification and phylogenetic analysis of the rickettsiae. Similar gene structure was observed for the citrate synthase gene[8] and the ATP translocator.[9] Although the codon usage indicated a bias toward A or U residues from the codon usage of *E. coli*, particularly in the wobble position for individual codons, the control elements such as promoters and ribosome binding sites shared homology with those of *E. coli*.

Since the initial descriptions of the cloning and sequencing of the first rickettsial genes, a number of rickettsial genes encoding enzymes, antigens, rRNAs, and proteins with a variety of different functions have been cloned and sequenced. Approximately 150 entries in the August 1996 version of GenBank can be accessed using Rickettsia as a search word. The development of PCR and rapid and automated sequencing methods have greatly facilitated sequencing of rickettsial genes. Large blocks of rickettsial genes are currently being sequenced. With the genome size of most rickettsiae being between one-third and one-half that of *E. coli*, it is likely that the entire genome of a rickettsia (and those of other rickettsiae) will be sequenced. Perhaps using this approach genes may be identified (or lack of genes) that explain the obligate intracellular nature of these organisms.

2.1. Transcription Initiation

As the transcription and translation of rickettsial genes are key steps in their expression in both *E. coli* and the rickettsiae themselves, these processes have been central to the study of rickettsial gene expression. One of the first descriptions of a rickettsial promoter was for the 17-kDa antigen gene. Putative promoter sequences for the 17-kDa antigen gene with homology to both −35 (TTTACA) and −10 (TATACT) promoter regions of *E. coli* were identified. Subcloning of these promoter regions into an *E. coli* promoter assay vector revealed that they resulted in transcription initiation of the β-galactosidase reporter gene with an efficiency of about one-half that of the *E. coli lac* promoter.[10] To determine if these regions with homology to the *E. coli* −10 and −35 promoter sequences function in the rickettsiae as they do in *E. coli*, the site of transcription initiation was mapped. The site of transcription initiation for the 17-kDa antigen gene was shown to be an adenine residue nine bases downstream of the −10 region. Similar sequences with equivalent levels of homology to the *E. coli* −35 and −10 promoter regions have been identified upstream from most of the rickettsial genes that have been subsequently sequenced. Consensus −35 and −10 regions can be derived by alignment of putative promoter regions identified from gene sequences

TTgacA   16N   TATANT
  -35   -10

FIGURE 1. Consensus promoter sequence compiled from rickettsial genes. Promoter sequences (identified as such in a GenBank entry) from genes for members of the genus *Rickettsia* were aligned for optimal homology. When multiple versions of genes are available (e.g., 17-kDa antigen gene), only one representative spotted fever group or one typhus group rickettsia was utilized. Positions conserved greater than 80% are capitalized, positions conserved 50–80% are in lowercase. Positions conserved less than 50% are indicated as N.

of both spotted fever and typhus group rickettsiae (Fig. 1). These results indicate that rickettsial promoters, at least the ones that have been cloned, appear to share both sequence homology and spacing between -35 and -10 regions of *E. coli* promoters. It is possible that the process of cloning rickettsial genes into *E. coli* and screening for expression of the gene product results in selection of rickettsial genes whose promoters are functional in *E. coli*. For that reason these promoters are likely to share homology with the consensus promoter sequences for *E. coli*.

Researchers have recently begun characterization of the RNA polymerase enzyme of *R. prowazekii*. The enzyme was shown to consist of the same types of subunits found in other bacterial RNA polymerases (α, β, β', and σ).[11] The RNA polymerase activity was found to be predominantly associated with the *R. prowazekii* cell membrane. This activity dissociated from the membrane with increasing salt concentration. The σ-factor subunit of the *R. prowazekii* RNA polymerase has been cloned and sequenced, yielding a 635-amino-acid protein with a calculated molecular size of 73 kDa.[12] The amino acid sequence has extensive homology with other bacterial RNA polymerase σ-factors. The σ-factor subunit expressed from *E. coli* minicells migrated at 85 kDa on SDS-PAGE, comigrating with the *R. prowazekii* enzyme. It has been speculated that the low ratio of RNA polymerase σ-factor to holoenzyme may result in a lower rate of transcription than for other bacteria and may serve to explain the slow growth rate of *R. prowazekii*.[13] It has also been shown that the rickettsial RNA polymerase was less efficient at transcribing $d(G + C)$-rich templates than the *E. coli* enzyme.[14] These findings taken together suggest that the rickettsial RNA polymerase subunit has similar composition and function to the enzyme from *E. coli*. Most and possibly all rickettsial promoters appear to function in *E. coli* as witnessed by transcription of cloned rickettsial genes. However, the rickettsial enzyme may have evolved to favor transcription of the $(A + T)$-rich genome of the rickettsiae and may explain (at least in part) the long generation time (10–12 hr) of these organisms.

2.2. Translation Initiation

Translation initiation in *E. coli* has been associated with a sequence immediately upstream of the start codon on the mRNA that has homology with the 3' end of the 16 S rRNA. This ribosome binding sequence, or Shine–Dalgarno sequence, presumably base pairs with the complementary sequence on the 16 S rRNA. It has been suggested that ribosome binding sites with the greatest ability to base pair with the 16 S rRNA initiate translation with the greatest efficiency. A consensus sequence has been proposed for *E. coli* based on the compilation of large numbers of ribosome binding sites. Polypurine-rich sequences have been found 4 to 14 bases upstream of the start codon for most rickettsial genes that have been sequenced. It is likely that these polypurine-rich sequences found in the rickettsiae function in a manner analogous to the ribosome binding sites of *E. coli*.

2.3. Posttranslational Modification

Processing of rickettsial proteins has been well characterized for two proteins of *R. rickettsii*. In the first case the 17-kDa antigen was shown to be lipid modified at least when expressed in *E. coli*. The deduced amino acid sequence for this protein was shown to contain the residues Leu-Gln-Ala-Cys at positions 17–20, a sequence that shares homology with the lipid modification sequence for lipoproteins of *E. coli*.[10] To determine if this was in fact a site for lipid addition, it was shown that the protein as expressed in *E. coli* was labeled with both [3H]palmitate and [3H]glycerol, compounds known to be sources of lipid moieties found in other bacterial lipoproteins. As lipid modification had previously been shown for other bacteria to occur at sequences similar to residues 17–20, *in vitro* mutagenesis was used to modify the cysteine codon at position 20 within that sequence. On altering Cys-20 to a glycine residue, lipid addition no longer occurred, although the 17-kDa protein was still expressed at similar levels to the wild-type version of the gene. These results suggest that the mechanism by which the cloned 17-kDa antigen is modified is similar to that for other bacterial lipoproteins.

A model for lipid modification of the 17-kDa antigen of *R. rickettsii* has been described.[15] In that model (Fig. 2) Cys-20 is first modified through a thioester linkage with glycerol (step 1). The newly attached glycerol is then modified through the addition of fatty acids at two of the three carbons (step 2). Like other bacterial lipoproteins, the glycerylcysteine-linked fatty acids most likely reflect the overall fatty acid composition of the cell. For *R. rickettsii* this has previously been shown to be octadecenoic acid ($C_{18:1}$) and palmitic acid ($C_{16:0}$).[16] Assuming the 17-kDa antigen of *R. rickettsii* is lipid-modified like other bacterial lipoproteins, then the fatty acids represented by R_1 and R_2 in Fig. 2 would most likely be octadecenoic or palmitic acid. After transport to the membrane, the lipid-modified 17-kDa antigen would then be cleaved at Cys-20 by a rickettsial signal peptidase (step 3). The newly created amino-terminus is then further modified by the addition of another lipid moiety, resulting in the mature protein. This model is based on the assumption that the expression of the cloned 17-kDa antigen in *E. coli* is identical in *R. rickettsii*. Technical obstacles have prevented experiments to demonstrate lipid modification of the 17-kDa antigen directly in *R. rickettsii*.

A second example of posttranslational processing can be found with the

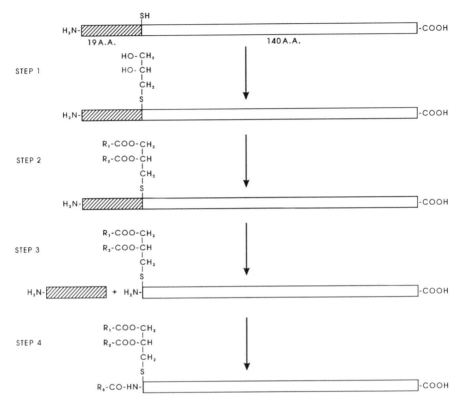

FIGURE 2. Model of lipid modification and cleavage of the 17-kDa antigen of *Rickettsia rickettsii*. Details of the model are described in the text. R_1, R_2, and R_3 represent fatty acid moieties. Reproduced with permission from *Rickettsiology: Current Issues and Perspectives*, K. Hechemy, D. Paretsky, D. H. Walker, and L. P. Mallavia, eds., Ann. N.Y. Acad. Sci. **590**:331.

rOmpB proteins apparently found in all members of the genus *Rickettsia*. The rOmpB is translated as a precursor protein that is then cleaved 32 kDa from the carboxy-terminus to yield a mature protein of approximately 120 kDa.[17,18] Both peptides appear to be associated on the surface of the rickettsiae as determined by coprecipitation with a monoclonal antibody specific for the 120-kDa portion of the protein. This protein may play a role in the virulence of *R. rickettsii*, as an avirulent mutant has reduced ability to cleave the precursor protein. This same mutant does not display the rOmpB protein (in any form) on the surface, unlike members of both the typhus and spotted fever group rickettsiae. These results suggest that cleavage of the precursor protein is an integral part of localization of rOmpB to the outer membrane.[18]

TABLE I
New Rickettsial Pathogens Identified with the Aid
of Molecular Biology Techniques

Organism	Reference
Bartonella henselae	21, 22
Ehrlichia chaffeensis	26
Human granulocytic ehrlichiosis agent (HGE)	28
Rickettsia felis	32
Rickettsia massiliae	42
Rickettsia japonica	43
Rickettsia helvetica	44

3. IDENTIFICATION OF NEW SPECIES

In the past 10 years a number of new rickettsial species have been identified and taxonomically defined (Table I). Almost invariably the methods used to define these new species include some type of genetic comparison with existing species. The most frequently utilized methods include DNA:DNA hybridizations and nucleotide sequence analysis of well-characterized genes such as the 16 S rRNA gene or the 17-kDa antigen gene. However, the application of DNA:DNA hybridizations, the recognized genetic technique to define new bacterial species, to the rickettsiae is often problematic. Many of the rickettsiae (e.g., *Orientia tsutsugamushi, Ehrlichia* species) are difficult to completely purify away from host cell debris and organelles, often a source of contaminating host cell DNA. Still other rickettsiae remain uncultivatable in an *in vitro* system. Accordingly, the task of pairwise DNA:DNA hybridizations between all of the rickettsiae becomes an impractical project that necessitates an alternate approach. The most frequently utilized technique is PCR amplification of the 16 S rRNA gene coupled with sequencing to define genetic divergence. The approach avoids the need to cultivate and purify the organisms away from host cell material, allowing comparisons of rickettsiae that are easily cultivated with those that have not yet been cultivated outside of natural or experimental animal hosts.

3.1. Bacillary Angiomatosis / Cat-Scratch Disease

Bacillary angiomatosis is a disease seen primarily among AIDS patients and manifests as lesions of the skin and visceral organs. These lesions are typified by vascular proliferation. Although bacillary angiomatosis was first described in 1983 by Stoler *et al.*,[19] the etiology remained unclear until 1990. Using broad-range PCR primers, Relman *et al.* were able to amplify bacterial 16 S rRNA gene sequences from the skin lesions of four patients with bacillary angiomatosis.[20] These broad-range primers are derived from highly conserved regions of the bacterial 16 S rRNA gene. When used for PCR amplification they will amplify fragments of the 16 S rRNA gene from virtually any bacterium. The sequences from these lesions were shown to belong to a bacterium related to but not identical to *Bartonella quintana*. This organism was isolated and subsequently shown to be a new species, *B. henselae*.[21,22]

Subsequent to this discovery, the same organism (*B. henselae*) was shown to be the central agent causing cat-scratch disease. Although a number of different laboratory methodologies were utilized to establish *B. henselae* as the etiologic agent of cat-scratch disease, again, broad-range PCR coupled with sequencing played an important role. *B. henselae* 16 S rRNA gene sequences were detected in the skin test antigens that had been used for many years for the diagnosis of cat-scratch disease.[23] No other bacterial 16 S rRNA gene sequences were detected indicating that the antigenic components of *B. henselae* are likely responsible for eliciting the delayed hypersensitivity that is observed with the cat-scratch disease skin test. Thus, bacteria that have been causing cat-scratch disease, which was originally described over 40 years ago,[24] have avoided detection and identification until the advent of modern molecular techniques.

3.2. Ehrlichiosis

In 1987, Maeda *et al.* described a patient with history of a tick bite suffering from symptoms similar to those seen in Rocky Mountain spotted fever.[25] Electron microscopy revealed intracellular inclusions, or morulae, characteristic of the rickettsiae belonging to the genus *Ehrlichia*. It was speculated that *E. canis* was the causative agent of this disease based on serologic reactivity of patient serum with *E. canis* antigen.[25] However, unique 16 S rRNA gene sequences were later identified by broad-range PCR amplification of the bacteria in the blood of two patients with ehrlichiosis. The sequences obtained from each of the two patients with ehrlichiosis were shown to be identical, but different from all existing species of *Ehrlichia*.[26] This same sequence was also identified in an isolate subsequently made from one of these two patients.[27] The name *E. chaffeensis* was proposed for this organism and PCR-based techniques remain the most reliable means to identify this rickettsia and differentiate it from the other members of this genus.[26]

The technique of broad-range PCR coupled with sequencing was also used for resolving the etiology of another form of ehrlichiosis, human granulocytic ehrlichiosis (HGE). In the initial description of HGE, three fatal cases occurred in a cluster of patients suffering from ehrlichiosis-like symptoms.[28] However, the morulae that are characteristic of ehrlichial infection were observed in the patients' granulocytes rather than in the monocytes as in cases of infection with *E. chaffeensis*. The organism was identified by amplifying bacterial 16 S rRNA gene fragments and sequencing the resulting PCR product. The resulting sequence was compared to 16 S rRNA gene sequences in the GenBank database. It was concluded that these patients were infected with a granulotropic ehrlichia either identical or closely related to E. phagocytophila.[28] This agent had never previously been associated with human disease before and had not at the time been isolated in the laboratory. Again, the powerful technique of broad-range PCR was used to amplify the E. phagocytophila-like 16 S rRNA gene sequences directly from the blood of infected patients.

3.3. ELB Agent

Another example of how molecular techniques can be used to detect and identify new rickettsial agents is seen in the case of the "ELB agent." A previously undescribed and uncultivatable rickettsial pathogen was detected in cat fleas by Azad et al. using PCR with conserved primers for the 17-kDa antigen gene and the citrate synthase gene.[29] Sequence analysis revealed that this agent had similarities to both the typhus and spotted fever group rickettsiae. Transovarial transmission of the ELB agent in fleas was demonstrated and the agent has been detected in a patient who had been diagnosed with murine typhus.[30] The ELB agent has subsequently been cultivated in tissue culture systems,[31] and the name "Rickettsia felis" has been proposed for this rickettsia.[32]

4. EPIDEMIOLOGY AND NATURAL HISTORY OF RICKETTSIAL DISEASES

Because of the role of vectors and potential animal reservoirs in transmission of the rickettsiae, molecular techniques have also been employed to identify sources and patterns of disease transmission. Again, PCR has been extensively used to detect various rickettsiae in potential vectors and animal reservoirs. Two examples of how animal reservoirs and vectors have been implicated in the transmission of diseases follow.

4.1. Ehrlichiosis

Since the first description of human monocytic ehrlichiosis in the United States and the identification of E. chaffeensis as the agent responsible, much effort has focused on identifying natural reservoirs and vectors for this agent. Because the process of isolating E. chaffeensis appears to be difficult and inefficient, PCR amplification of E. chaffeensis-specific sequences has proven valuable in identifying potential animal reservoirs and arthropod vectors.

An E. chaffeensis-specific PCR assay targeting a segment of the 16 S rRNA gene has been developed and used to detect the organisms in blood from infected patients. In that report, *E. chaffeensis* was detected in a single *Dermacentor variabilis* tick collected from an opossum found at the same location (Fort Chaffee Army Base, Arkansas) as two of the patients from whose blood *E. chaffeensis* was amplified.[33] In a subsequent report using large numbers of tick pools collected from multiple sites, *E. chaffeensis* was found only in *Amblyomma americanum* ticks and not *Dermacentor* species.[34] It had been speculated that *A. americanum* was a vector of human monocytic ehrlichiosis because the geographic distribution of the disease is remarkably similar to the range of this tick. A potential animal reservoir was identified when white-tailed deer were shown to have antibodies to *Ehrlichia* species.[35] Further support for a role of deer as a significant harborer of *E. chaffeensis* was garnered when transient rickettsemia in white-tailed deer was demonstrated using PCR.[36] Again, PCR was used to show experimental transmission of *E. chaffeensis* among white-tailed deer by *A. americanum* nymphal and adult ticks.[37] Thus, *A. americanum* is most likely the major vector responsible for transmission of *E. chaffeensis* to humans.

The identification of a potential tick vector for HGE has benefited from molecular techniques as well. Phylogenetic analysis of the agent of HGE, *E. equi*, and *E. phagocytophila* indicates that these agents are either very closely related or identical and form a single group. Accordingly, the role of the hosts and vectors of these agents has been investigated to define a natural reservoir of HGE. Recently, Pancholi *et al.* detected DNA from member(s) of the HGE group of ehrlichia in 7 of 68 *Ixodes dammini* ticks collected in Wisconsin.[38] *Ixodes* species are the most likely vector of the agent of HGE. Transmission and passage of the agent of HGE in horses has been demonstrated.[39] Thus, both phylogenetic analysis of the 16 S rRNA gene as well as the host range suggest that the agent of HGE may be a strain of *E. equi* and both are closely related (or identical) to *E. phagocytophila*.

4.2. Bartonella

The association of cats with cat-scratch disease has been known since the disease was first described by Debré *et al.* in 1950.[24] Since the association of *B. henselae* with cat-scratch disease and the association of both *B. henselae* and *B. quintana* with bacillary angiomatosis, much focus has been directed at understanding the role of cats in transmission of these agents. The presence of *B. henselae* in naturally infected cats has been documented by both isolation and PCR.[40] PCR was used to detect *B. henselae* in the cat flea (*Ctenocephalides felis*), providing the first evidence that fleas may play an important role in transmission of *Bartonella*.[40] Experimental transmission of *b. henselae* from infected cats to specific-pathogen-free kittens was demonstrated and utilized PCR to detect the organisms in infected fleas.[41] However, no cat-to-cat transmission was observed in the absence of fleas, suggesting that fleas are the primary means of transmission among cats. It is likely that transmission to humans occurs via a scratch from a cat whose claw is contaminated with feces or debris from infected fleas or cat blood.

5. SUMMARY/CONCLUSION

With the identification of rickettsiae and rickettsia-like pathogens as agents of emerging disease, there is increased importance in studying these organisms. The study of the rickettsiae has been greatly fostered by the application of molecular techniques and tasks once insurmountable, now have become just another routine application of "gene-jockeying." The use of PCR has probably been the single greatest tool afforded the rickettsiologists. Single genes can now be amplified from just a few organisms allowing detection for diagnostic purposes and sources of genes for research purposes as well. The need to identify a single clone in an entire library of DNA from rickettsiae is slowly being reduced. As outlined in this brief review, applications of molecular techniques have impacted a variety of different areas of study ranging from diagnostics to the most basic research. Efforts are now under way to sequence the entire genome of certain rickettsiae. The gene sequences that may result from such projects may help unravel the mysteries surrounding the unique intracellular niche that the rickettsiae occupy. There is little doubt that molecular techniques and future refinements of these methodologies will be central to future studies encompassing a variety of disciplines in rickettsiology.

REFERENCES

1. Wood, D. O., Atkinson, W. H., Sikorski, R. S., and Winkler, H. H., 1983, Expression of the *Rickettsia prowazekii* citrate synthase gene in *Escherichia coli, J. Bacteriol.* **155:**412–416.
2. Krause, D. C., Winkler, H. H., and Wood, D. O., 1985, Cloning and expression of the *Rickettsia prowazekii* ADP/ATP translocator in *Escherichia coli, Proc. Natl. Acad. Sci. USA* **82:**3015–3019.
3. Oaks, E. V., Stover, C. K., and Rice, R. M., 1987, Molecular cloning and expression of *Rickettsia tsutsugamushi* genes for two major protein antigens in *Escherichia coli, Infect. Immun.* **55:**1156–1162.
4. Anderson, B. E., Regnery, R. L., Carlone, G. C., Tzianabos, T., McDade, J. E., Fu, Z. Y., and Bellini, W. J., 1987, Sequence analysis of the 17-kilodalton-antigen gene from *Rickettsia rickettsii, J. Bacteriol.* **169:**2385–2390.
5. McDonald, G. A., Anacker, R. L., and Garjian, K., 1987, Cloned gene of *Rickettsia rickettsii* surface antigen: Candidate vaccine for Rocky Mountain spotted fever, *Science* **235:**83–85.
6. Anderson, B. E., and Tzianabos, T., 1989, Comparative sequence analysis of a genus-common rickettsial antigen gene, *J. Bacteriol.* **171:**5199–5201.
7. Winkler, H. H., and Wood, D. O., 1988, Codon usage in selected AT-rich bacteria, *Biochimie* **70:**977–986.
8. Wood, D. O., Williamson, L. R., Winkler, H. H., and Krause, D. C., 1987, Nucleotide sequence of the *Rickettsia prowazekii* citrate synthase gene, *J. Bacteriol.* **169:**3564–3572.

9. Williamson, L. R., Plano, G. V., Winkler, H. H., Krause, D. C., and Wood, D. O., 1989, Nucleotide sequence on the *Rickettsia prowazekii* ATP/ADP translocase-encoding gene, *Gene* **80:**269–278.

10. Anderson, B. E., Baumstark, B. R., and Bellini, W. J., 1988, Expression of the gene encoding the 17-kilodalton-antigen from *Rickettsia rickettsii:* Transcription and posttranslational modification, *J. Bacteriol.* **170:**4493–4500.

11. Ding, H. F., and Winkler, H. H., 1990, Purification and partial characterization of the DNA-dependent RNA polymerase from *Rickettsia prowazekii, J. Bacteriol.* **172:**5624–5630.

12. Marks, G. L., Winkler, H. H., and Wood, D. O., 1992, Isolation and characterization of the gene coding for the major sigma factor of *Rickettsia prowazekii* DNA-dependent RNA polymerase, *Gene* **121:**155–160.

13. Ding, H. F., and Winkler, H. H., 1994, The molar ratio of sigma 73 to core polymerase in the obligate intracellular bacterium, *Rickettsia prowazekii, Mol. Microbiol.* **11:**869–873.

14. Ding, H. F., and Winkler, H. H., 1993, Characterization of the DNA-melting function of the *Rickettsia prowazekii* RNA polymerase, *J. Biol. Chem.* **268:**3897–3902.

15. Anderson, B. E., 1989, The 17-kilodalton antigens of spotted fever and typhus group rickettsiae, *Ann. N.Y. Acad. Sci.* **590:**326–333.

16. Tzianabos, T., Moss, C. W., and McDade, J. E., 1981, Fatty acid composition of rickettsiae, *J. Clin. Microbiol.* **13:**603–605.

17. Gilmore, R. D., Jr., Cieplak, W., Jr., Policastro, P. F., and Hackstadt, T., 1991, The 120 kilodalton outer membrane protein (rOmp B) of *Rickettsia rickettsii* is encoded by an unusually long open reading frame: Evidence for protein processing from a large precursor, *Mol. Microbiol.* **5:**2361–2370.

18. Hackstadt, T., Messer, R., Cieplak, W., and Peacock, M. G., 1992, Evidence for proteolytic cleavage of the 120-kilodalton outer membrane protein of rickettsiae: Identification of an avirulent mutant deficient in processing, *Infect. Immun.* **60:**159–165.

19. Stoler, M. H., Bonfiglio, T. A., Steigbigel, R. T., and Pereira, M., 1983, An atypical subcutaneous infection associated with acquired immune deficiency syndrome, *Am. J. Clin. Pathol.* **80:**714–718.

20. Relman, D. A., Loutit, J. S., Schmidt, T. M., Falkow, S., and Tompkins, L. S., 1990, The agent of bacillary angiomatosis: An approach to the identification of uncultured pathogens, *N. Engl. J. Med.* **323:**1573–1580.

21. Regnery, R. L., Anderson, B. E., Rodriquez, M., Jones, D. C., and Carr, J. H., 1992, Characterization of a novel *Rochalimaea* species, *R. henselae* sp. nov., isolated from blood of a febrile, human immunodeficiency virus-positive patient, *J. Clin. Microbiol.* **30:**265–274.

22. Welch, D. F., Pickett, D. A., Slater, L. N., Steigerwalt, A. G., and Brenner, D. J., 1992, *Rochalimaea henselae* sp. nov., a cause of septicemia, bacillary angiomatosis, and parenchyman bacillary peliosis, *J. Clin. Microbiol.* **30:**275–280.

23. Anderson, B., Kelly, C., Threlkel, R., and Edwards, K., 1993, Detection of *Rochalimaea henselae* in cat-scratch disease skin-test antigens, *J. Infect. Dis.* **168:**1034–1036.

24. Debré, R., Lamy, M., Jammet, M.-L., Costil, L., and Mozziconacci, P., 1950, La maladie des griffes de chat, *Bull. Mem. Soc. Med. Hop. Paris* **66:**76–79.

25. Maeda, K., Markowitz, N., Hawley, R. C., Ristic, M., Cox, D., and McDade, J. E., 1987, Human infection with *Ehrlichia canis,* a leukocytic rickettsia, *N. Engl. J. Med.* **316:**853–856.

26. Anderson, B. E., Dawson, J., Jones, D., and Wilson, K., 1991, *Ehrlichia chaffeensis,* a new species associated with human ehrlichiosis, *J. Clin. Microbiol.* **29:**2838–2842.

27. Dawson, J., Anderson, B. E., Fishbein, D. B., Sanchez, J. L., Goldsmith, C. S., Wilson, K. H., and Duntley, C. W., 1991, Isolation and characterization of an *Ehrlichia* sp. from a patient diagnosed with human ehrlichiosis, *J. Clin. Microbiol.* **29:**2741–2745.

28. Chen, S. M., Dumler, J. S., Bakken, J. S., and Walker, D. H., 1994, Identification of a granulocytic *Ehrlichia* species as the etiologic agent of human disease, *J. Clin. Microbiol.* **32:**589–595.
29. Azad, A. F., Sacci, J. B., Jr., Nelson, W. M., Dasch, G. A., Schmidtmann, E. T., and Carl, M., 1992, Genetic characterization and transovarial transmission of a typhus-like rickettsia found in cat fleas, *Proc. Natl. Acad. Sci. USA* **89:**43–46.
30. Schriefer, M. E., Sacci, J. B., Jr., Dumler, J. S., Bullen, M. G., and Azad, A. F., 1994, Identification of a novel rickettsial infection in a patient diagnosed with murine typhus, *J. Clin. Microbiol.* **32:**949–954.
31. Radulovic, S., Higgins, J. A., Jaworski, D. C., Dasch, G. A., and Azad, A. F., 1995, Isolation, cultivation, and partial characterization of the ELB agent associated with cat fleas, *Infect. Immun.* **63:**4826–4829.
32. Higgins, J. A., Radulovic, S., Schreifer, M. E., and Azad, A. F., 1996, *Rickettsia felis:* A new species of pathogenic rickettsia isolated from cat fleas, *J. Clin. Microbiol.* **34:**671–674.
33. Anderson, B. E., Sumner, J. W., Dawson, J. E., Tzianabos, T., Greene, C. R., Olson, J. G., Fishbein, D. F., Olsen-Rasmussen, M., Holloway, B. P., George, E. H., and Azad, A. F., 1992, Detection of the etiologic agent of human ehrlichiosis by polymerase chain reaction, *J. Clin. Microbiol.* **30:**775–780.
34. Anderson, B., Sims, K., Olson, J., Childs, J., Piesman, J., Happ, C., Maupin, G., and Johnson, B., 1993, *Amblyomma americanum:* A potential vector of human ehrlichiosis, *Am. J. Trop. Med. Hyg.* **49(2):**239–244.
35. Dawson, J. E., Childs, J. E., Biggie, K. L., Moore, C., Stallknecht, D., Shaddock, J., Bouseman, J., Hofmeister, E., and Olson, J. G., 1994, White-tailed deer as a potential reservoir of *Ehrlichia* spp., *J. Wildl. Dis.* **30:**162–168.
36. Dawson, J. E., Stallknecht, D. E., Howerth, E. W., Warner, C., Biggie, K., Davidson, W. R., Lockhart, J. M., Nettles, V. F., Olson, J. G., and Childs, J. E., 1994, Susceptibility of white-tailed deer (*Odocoileus virginianus*) to infection with *Ehrlichia chaffeensis,* the etiologic agent of human ehrlichiosis, *J. Clin. Microbiol.* **32:**2725–2728.
37. Ewing, S. A., Dawson, J. E., Koca, A. A., Barker, R. W., Warner, C. K., Panciera, R. J., Fox, J. C., Kocan, K. M., and Blouin, E. F., 1995, Experimental transmission of *Ehrlichia chaffeensis (Rickettsiales: Ehrlichieae)* among white-tailed deer by *Amblyomma americanum (Acari: Ixodidae), J. Med. Entomol.* **32:**368–374.
38. Pancholi, P., Kolbert, C. P., Mitchell, P. D., Reed, K. D., Jr., Dumler, J. S., Bakken, J. S., Telford, S. R., 3rd, and Persin, D. H., 1995, *Ixodes dammini as a potential vector of human granulocytic ehrlichiosis, J. Infect. Dis.* **172:**1007–1012.
39. Madigan, J. E., Richter, P. J., Jr., Kimsey, R. B., Barlough, J. E., Bakken, J. S., and Dumler, J. S., 1995, Transmission and passage in horses of the agent of human granulocytic ehrlichiosis, *J. Infect. Dis.* **172:**1141–1144.
40. Koehler, J. E., Glaser, C. E., and Tappero, J. W., 1994, *Rochalimaea henselae* infection: New zoonosis with the domestic cat as reservoir, *J. Am. Med. Assoc.* **271:**531–535.
41. Chomel, B. B., Kasten, R. W., Floyd-Hawkins, K., Chi, B., Yamamoto, K., Roberts-Wilson, J., Gurfield, A. N., Abbott, R. C., Pedersen, N. C., and Koehler, J. E., 1996, Experimental transmission of *Bartonella henselae* by the cat flea, *J. Clin. Microbiol.* **34:**1952–1956.
42. Beati, L., and Raoult, D., 1993, *Rickettsia massiliae* sp. nov., a new spotted fever group rickettsia, *Int. J. Syst. Bacteriol.* **43:**839–840.
43. Uchida, T., Uchiyama, T., Kumano, K., and Walker, D. H., 1992, *Rickettsia japonica* sp. nov., the etiologic agent of spotted fever group rickettsiosis in Japan, *Int. J. Syst. Bacteriol.* **42:**303–305.
44. Beati, L., Peter, O., Burgdorfer, W., Aeschlimann, A., and Raoult, D., 1993, Confirmation that *Rickettsia helvetica* sp. nov. is a distinct species of the spotted fever group rickettsiae, *Int. J. Syst. Bacteriol.* **43:**521–526.

2

Immunity to Rickettsiae (Redux)

THOMAS R. JERRELLS

1. INTRODUCTION

As intracellular bacteria, the group of microorganisms collectively and commonly known as the rickettsiae have established a unique niche, one believed to have given these organisms a selective advantage. It is generally believed that the intracellular bacteria have established the ability to replicate in the host cell so as to evade the immune system. This clearly works as a protective mechanism to evade antibody, which cannot enter the host cell. It would seem logical that the cell-mediated immune system evolved in response to intracellular parasites. As will be discussed in later sections, the primary (but not exclusive) mediator of immunity to obligate intracellular bacteria is the cell-mediated immune response and especially the delayed-hypersensitivity response.

The members of the group of intracellular bacteria to be discussed in this chapter are diverse and dissimilar in their biological characteristics. In spite of these differences, some general conclusions can be made regarding the immune response to these organisms. The purposes of this discussion are threefold: (1) to review these general concepts of immunity, (2) to provide some new information obtained from studies in this and other laboratories, and (3) to discuss the possible immunologic differences associated with the biologic differences in the members of this group. In this chapter I will lump together members of the genera *Rickettsia*, *Orientia (Rickettsia tsutsugamushi)*, *Ehrlichia*, and *Coxiella*. It is of course naive to assume that the immune response to these organisms is identical. However, it is

THOMAS R. JERRELLS • Department of Pharmaceutical Sciences, College of Pharmacy, Washington State University, Pullman, Washington 99164-6510.

Rickettsial Infection and Immunity, edited by Anderson *et al.* Plenum Press, New York, 1997.

very likely that the immune response to this diverse group shares many similarities. As complex as the immune system is, it is simpler than the biology of the obligate intracellular microorganisms.

The reader is urged to refer to an earlier review of the literature on the immune responses to the rickettsiae[1] for a detailed description of experiments that have so elegantly set the stage for the current state of knowledge about these organisms.

2. IMMUNE RESPONSES TO INTRACELLULAR ORGANISMS

Much information has been published that addresses the mechanisms of immunity to intracellular bacteria, and the reader should consult a recent review by Kaufmann[2] for details that will not be presented in this chapter.

It is generally believed that immunity to obligate intracellular bacteria is a T-cell-dependent response regardless of whether the response is mediated by a cell-mediated immunity or an antibody.[1] Antigen-specific T cells can be separated phenotypically into two distinct groups: helper T cells and cytotoxic T cells. The former express the CD4 molecule on their surface, and the latter characteristically express CD8 molecules. Both CD4 and CD8 molecules are involved in the interaction between the T cell and antigen-presenting cells, which determine the molecular nature of the antigenic epitopes recognized by each of these cell types.

Recent study findings have shown that the T-cell-dependent immune responses in rodents—and, for the most part, in human beings—that are mediated by the $CD4^+$ T cell can be separated biologically into at least two effector cell populations on the basis of their characteristic cytokine production.[3-5] Briefly, the TH1 $CD4^+$ T cell uniquely produces interleukin 2 and gamma interferon (IFN-γ). This is the helper T cell that is involved in the so-called inflammatory cell-mediated immune response, which is the quintessential delayed-hypersensitivity response. The delayed-type hypersensitivity response depends absolutely on the production of IFN-γ and therefore depends on the TH1 T cell.[6] The TH1 T cell also is involved in the production of T-dependent IgG2 antibodies, again through the action of IFN-γ.[5] Thus, the concept of a cellular and humoral immune response dichotomy is somewhat obsolete. The TH1 T cell modulates the production and function of the TH2 T cell through the production of IFN-γ by obscure mechanisms. It is noteworthy that the natural killer cell also produces IFN-γ, and it is very likely that this cell is of critical importance in either the development of a TH1 immune response to rickettsiae or the mediation of immunity to these organisms, as has been suggested for immunity to *Listeria monocytogenes* and *Salmonella typhimurium*.[7]

Of interest is the recent finding that development of the antigen-specific TH1 T cell from naive helper T cells depends on production of interleukin 12, which is produced by the macrophage. This cytokine also interacts with the natural killer cell to induce IFN-γ production, acts as a growth factor for activated T and natural killer cells, enhances the lytic activity of natural killer and lymphokine-activated killer cells, and facilitates antigen-specific cytotoxic T-cell responses. In addition, this cytokine acts as a costimulus for maximal production of IFN-γ by the activated TH1 T cell.[8-13] It is logical and compelling to suggest that interleukin 12 would be of importance for immunity to rickettsiae, but no data exist to support this conclusion.

In contrast to the cytokine profile of the TH1 T cell, the TH2 T cell produces interleukin 4, 5, 10, and 13 uniquely. It is also involved in the production of IgE, IgA, and IgG1 antibodies[4,5] and is thus the classic helper T cell. TH2 T cells cross-regulate the production and function of the TH1 T cell in an analogous manner as the TH1 T cell interacts with the TH2 T cell but through the production and action of interleukin 4 and 10 on the TH1 T cell.[14,15] Although the TH2 T cell probably is not involved directly in immunity to intracellular bacteria as are TH1 T and natural killer cells, the TH2 T-cell function may be critical for immune resistance to a secondary infection with the rickettsial organisms (e.g., *Coxiella burnetii*) that infect through the mucosa. Furthermore, production of a predominant TH2 cellular response is associated with well-characterized genetic susceptibilities to several intracellular organisms[14] and may be involved in the genetic susceptibility of some strains of mice to rickettsial infection.[16] It would be premature to exclude a TH2 immune response to the rickettsiae without further studies.

Both the TH1 and TH2 helper T cells are restricted to antigen recognition by the class II major histocompatibility complex (MHC) on the antigen-presenting cell. Depending on the stage of the immune response that is examined, the function of antigen presentation can be carried out by the dendritic cell or macrophage (primary responses), the follicular dendritic cell in the lymph nodes (also a primary immune response), or the B cell (secondary or anamnestic immune responses). It is important to note that the expression of a secondary or memory immune response occurs optimally when antigen is presented to memory T cells by antigen-specific B cells. Thus, development of antibody-producing cells may not be critical for the mediation of antirickettsial immunity, but it certainly plays a role in the total response to these organisms.

Early work by Rollwagen *et al.*[17] suggested a role for the CD8+ T cell in immunity to rickettsiae, but no mechanism for the action of this cell has come forth. The CD8+ T cell is of critical importance in immunity to other intracellular bacteria such as *L. monocytogenes.*[18] The mechanism for immunity that is mediated by these cells is that of IFN-γ production, as well as perhaps through that of other cytokines such as tumor necrosis factor. The CD8+ T cell apparently is the major mediator of immunity to those intracellular bacteria that are capable of leaving the phagolysosome of the phagocytic cell and entering the cytoplasm.[19]

For this reason, it can be concluded that this cell is of primary importance in immunity to those rickettsiae that replicate in the cytoplasm of the host cell. This would include members of the genera *Rickettsia* and *Orientia* (*R. tsutsugamushi*). The reader is referred to later sections for further information on the role of CD8+ T cells in immune responses to rickettsiae.

T cells recognize antigen through a receptor on their surface, the T-cell receptor. The CD4+ and CD8+ T cells characteristically have a T-cell receptor composed of α/β chains. Another subset of T cells that have been shown to have a role in immunity to intracellular bacteria has a T-cell receptor that characteristically consists of γ/δ chains.[20,21] The mechanisms by which the T cells with a T-cell receptor composed of γ/δ chains mediate immunity are unknown. It also is not known whether they might play a role in immunity to rickettsiae.

3. ROLE OF THE T CELL IN IMMUNITY TO RICKETTSIAE

Results from a number of studies support the conclusion that immunity to the obligate intracellular bacteria is mediated by T cells. This evidence includes studies that were done with passive transfer of immune lymphocytes and cell populations enriched for T cells.[22-27] It is well known that T cells are required for antibody production, and it is possible that the transferred T cells simply enhance antibody production in the recipient. However, results of most experiments that have been done with passive transfer of immune sera show that this did not protect.[22] The reader is referred to a later section for information regarding the possible role of antibody in secondary resistance to infection with these organisms.

In vitro parameters of T-cell responsiveness to rickettsial antigens have also helped establish the importance of antigen-responsive T cells in rickettsial immunity. These studies have been done with the use of *in vitro* lymphocyte proliferation assays,[28-36] as well as with the production of T-cell-produced cytokines—especially IFN-γ.[37-41] Because these assays have been applied to a number of species, including human beings and nonhuman primates,[28-30,33] the conclusions drawn from these data can be easily generalized to indicate the importance of the antigen-responsive T cell in rickettsial immunity.

To determine the role of the CD4+ and CD8+ T cells in immunity to *O.* (*Rickettsia*) *tsutsugamushi*, preliminary experiments were done in this laboratory where each of these T-cell subpopulations was depleted with monoclonal antibodies before challenge of immune mice with the Karp strain of this rickettsia. Results of these studies are summarized in Table I. These data support the conclusion that both CD4 and CD8 cells are necessary for resistance to reinfection of immune mice. As suggested by others, it is likely that the CD8 T cell requires the CD4 cell to develop effector function fully. In other preliminary

TABLE I
Effect of T-Cell Subset Depletion on Mortality of Mice Immune to *Rickettsia tsutsugamushi* after a Secondary Challenge with Homologous Rickettsiae[a]

Treatment[b]	Experiment 1	Experiment 2
Anti-CD4[c]	0%(0/3)[d]	0%(0/3)
Anti-CD8[e]	0%(0/3)	33%(1/3)
Anti-CD4 and Anti-CD8	100%(3/3)[f]	100%(4/4)[g]
None	0%(0/6)	0%(0/4)

[a]Female C57Bl/6 mice were immunized with 1000 plaque-forming units of the Karp strain of *Rickettsia tsutsugamushi* 28 days before challenge with 10,000 plaque-forming units of the same strain of rickettsia.

[b]Each animal was given two doses of monoclonal antibody at 500 μg per dose. The first dose was administered 24 hr before the challenge; the second dose, 72 hr after the challenge infection.

[c]The GK 1.5 hybridoma was used to produce the anti-CD4 monoclonal antibody.

[d]Survival. Animals were monitored for 4 weeks after infection for mortality.

[e]The YTS 169 hybridoma was used to produce the anti-CD8 monoclonal antibody.

[f]Two animals died 12 days after infection; one died 14 days after infection.

[g]Two animals died 10 days after infection; one, 12 days after infection; and one, 16 days after infection.

experiments (data not shown), depletion of both T-cell subsets resulted in recurrence of rickettsial infection in these chronically infected animals. This finding supports the suggestion that CD4 and CD8 cells are necessary to control the chronic infections. The existence of antigen-responsive CD8$^+$ T cells was further supported by the demonstration of class I MHC-restricted CD8$^+$ cells that responded specifically to rickettsial antigens (Fig. 1). In these experiments antigen-specific cytotoxic activity was also demonstrated. This function had identical characteristics in terms of blocking with anti-CD8 and anti-class I MHC monoclonal antibodies. It is clear from these data that CD8 cells are important in immunity and easily demonstrable *in vitro*.

Together these data would support the conclusion that immunity to *O. (Rickettsia) tsutsugamushi* is mediated by both CD4$^+$ and CD8$^+$ T cells. However, which cell plays the more important role (if any role) and the mechanisms of immunity to intracellular bacteria mediated by each cell type are still unknown.

Recently, the interesting observation[42] was made that spotted-fever-group organisms that were unrelated serologically nevertheless had complete ability to cross-stimulate immune T cells. This cross-reaction was associated with the establishment of protective immunity of guinea pigs to challenge with the virulent *Rickettsia rickettsii*. These data support the suggestion that epitopes recognized by antibody may not be protective and in fact may not be on the same protein as the epitopes that stimulate T cells.

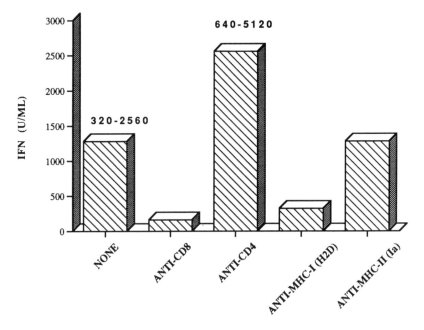

FIGURE 1. Gamma interferon production by CD8$^+$ spleen cells obtained from mice immune to *Rickettsia tsutsugamushi.* Spleen cells obtained from mice immunized with the Karp strain of *R. tsutsugamushi* were depleted of macrophages by adherence to plastic and cultured with macrophages obtained from bone marrow precursor cells in the presence of the indicated antibodies. Heat-killed rickettsiae were used as the antigen for stimulation. With the use of flow cytometric methods the bone marrow-derived macrophages were shown to express only class I major histocompatibility antigen. Numbers above the columns indicate the range of interferon units obtained with triplicate assays if applicable. Each point represents the mean of triplicate determinations.

4. ROLE OF ANTIBODY IN RICKETTSIAL IMMUNITY

Because immunoglobulin cannot enter infected cells, it has been argued that antibody does not mediate directly clearance of rickettsiae from the infected cell. This conclusion is supported by data that show that rickettsial infection can continue in the face of vigorous antibody production in athymic mice.[43-45] These data do not rule out a role for T-cell-dependent antibody, but passive transfer of immune sera that contained high levels of what must be presumed to be T-dependent antibodies has usually failed to protect against infection when given either before or after the infection. On the other hand, there are study results that show that pretreatment of rickettsiae with immune sera or monoclonal antibodies[46] prevents infection or reduces infectivity. From these data it could be suggested that antibody is important in *secondary* immune responses to these organisms by binding the organisms to prevent attachment or penetration of the host cell by the organisms, which is analogous to the function of secretory IgA in resistance to secondary infections of the gastrointestinal mucosa by *S. typhimurium*.[47] This could be related to antibody binding to a specific rickettsial ligand that interacts with the cell surface, or simply to steric hindrance of the organism's attachment to the surface. Further studies are required to confirm and expand these ideas.

5. ROLE OF CYTOKINES IN RICKETTSIAL IMMUNITY

The importance of IFN-γ in activation of macrophages has been well established in a number of model systems, including rickettsial systems.[39–41,48,49] This cytokine seems to be active against rickettsiae that reside in the cell cytoplasm, as well as against organisms (e.g., *C. burnetii* and *Ehrlichia* species) that remain in the phagolysosome. Data obtained from studies done in rodents with the use of a diverse group of facultative intracellular bacteria, such as *L. monocytogenes*, *S. typhimurium*, *Mycobacterium tuberculosis*, and *Legionella pneumophila*, support the conclusion that IFN-γ acts predominantly by inducing the production of reactive nitrogen intermediates, especially nitric oxide.[50,51] It is not known whether this mechanism is applicable to the rickettsiae, but some evidence exists[52] that supports this as a likely possibility. Care must be taken when generalizing these data because of the uniform failure to demonstrate the production of reactive nitrogen intermediates by cells obtained from human beings. IFN-γ inhibits the replication of diverse organisms, including *Rickettsia typhi*, *O. (Rickettsia) tsutsugamushi*, and *C. burnetii*, in cells other than macrophages.[53–57] Similar data have been generated by Szalay *et al.*[58] that show that growth of *L. monocytogenes* in hepatocytes is likewise inhibited by treatment of the cell with IFN-γ. It is still unclear whether this cytokine inhibits the growth of rickettsiae in endothelial cells, which are, after all, the most relevant cell to study. Regardless, it is a distinct possibility that IFN-γ, perhaps produced by natural killer cells, acts on the infected cell to inhibit the growth of rickettsiae by unknown mechanisms long enough to allow the immune system to catch up and ultimately clear the organisms.

Infection of macrophages with rickettsiae from the spotted fever and scrub typhus groups results in the stimulation of tumor necrosis factor (TNF) production by these cells.[59,60] It is interesting to note that these observations have been obtained with the use of cells from both mice and human beings. Results of studies to date show that the rickettsial organisms that have an outer lipopolysaccharide layer induce larger amounts of TNF.[59] It is interesting that purified lipopolysaccharide obtained from *C. burnetii* does not induce macrophages from mice to produce large levels of TNF (T. Jerrells and T. Hackstadt, unpublished data). This interesting difference in the lipopolysaccharide of these organisms

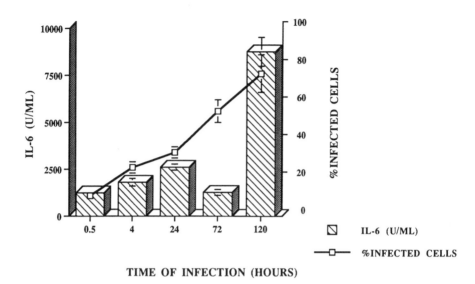

FIGURE 2. Production of interleukin 6 after infection of human umbilical endothelial cells with the Gilliam strain of *Rickettsia tsutsugamushi*. Endothelial cells were infected with the Gilliam strain of *R. tsutsugamushi* at a multiplicity of infection of 10. Culture supernatants were obtained at the indicated times and assayed for interleukin 6 with the use of a bioassay that was done with 7TD-1 cells. Each point represents the mean ± one standard deviation of triplicate cultures. The endothelial cells were recovered from the culture dishes at identical times with the use of trypsin, and the percentage of cells infected was determined after staining with Wright–Giemsa stain. Similar data were obtained when the supernatants were assayed for interleukin 8.

deserves further study. In a recent paper[61] it was reported that TNF activated murine macrophages to inhibit the replication of *O. (Rickettsia) tsutsugamushi*. However, the mechanisms of this inhibition are unclear. As would be expected, TNF and IFN-γ together were more effective than either alone for macrophage activation to inhibit rickettsial growth.

Infection of endothelial cells *in vitro* with *R. rickettsii*[62] or *O. (Rickettsia) tsutsugamushi* (Fig. 2) results in the production of cytokines, such as interleukin 6, 8, and 1, by the endothelial cell. These cytokines are important in the development of an inflammatory response at the site of infection and may well be critical for the ultimate clearance of the organisms. It is important to determine with the use of cellular or molecular techniques whether similar effects are noted *in vivo*. As noted above, it is critical to determine whether these phenomena are applicable to the vascular endothelial cell, as well as to extend the *in vitro* findings to *in vivo* systems.

6. IMMUNE RESPONSES AND CHRONIC INFECTIONS

An interesting observation that has not received much attention is the development of chronic rickettsial infections [*R. typhi* in human beings and *O. (Rickettsia) tsutsugamushi* in rodent models] in the face of a strong humoral and cellular immune response. It is clear that the establishment of a chronic infection does not involve large antigenic or phenotypic changes in the organisms. How the rickettsial organisms evade the immune system or establish an equilibrium with the immune system is unknown and would provide a very intersting topic for future research. It would seem at first glance that the idea that an equal rate of rickettsial replication and immune reactivity is established in chronic infections has merit, but the immune system adapts as it develops to provide for cells that are exquisitely specific for the antigenic epitope and the ability to respond very quickly when it recognizes antigen for the second time. This anamnestic response phenomenon has been studied for years and involves the development of memory T cells.

Another possibility that deserves further study is the suppression of aspects of the immune response, thus allowing the rickettsiae to survive in the host for long periods. There are data in the literature that demonstrate alterations in immune responses associated with rickettsial infections,[63-68] but very little is known about the specific mechanism(s) responsible for these changes.

Although the possibilities discussed above deserve further study and are interesting in their own right, it is more likely that the rickettsiae have discovered an immunologically privileged area to hide in. Again, much work is needed to probe this interesting question.

The possibility that antibiotic-resistant strains of *O. (Rickettsia) tsutsugamush*: have emerged in Asia further underscores the importance of the immune system and opens questions of why the immune system is not completely effective in protecting human beings from serious illness when infected with these organisms.

7. SUMMARY

The obligate intracellular bacteria that we call rickettsiae obviously share many biological traits with the other intracellular bacteria. For example, it has been shown that both *L. monocytogenes* and *R. rickettsii* move through the cell cytoplasm and from cell to cell by an identical technique of actin polymerization.[69] It is also clear that the intracellular bacteria share many aspects of the immune response, and this is not surprising. From the existing data it can be concluded that immunity to primary rickettsial infections characteristically is a TH1-type cellular immunity. Further, IFN-γ is probably the most important cytokine for clearance of the organisms, although clearly not the only cytokine of importance. This statement is apparently true for the members of the genera *Rickettsia* and *Ehrlichia*, as well as for *C. burnetii*. Infection with these organisms has been noted to be associated with an immunosuppression, and this observation is largely unstudied. Perhaps the more important issue is the mechanisms that these organisms use to evade the immune system rather than the mechanisms of immunity that clear the organisms.

Probably the most compelling basic question that can be approached with the use of rickettsial model systems is how the immune system interacts with the vascular endothelium in terms of antigen presentation by the endothelial cell and T-cell interactions with the infected endothelial cell. It is hoped this review of the immune response to rickettsiae will inspire scientists to tackle these and other questions.

ACKNOWLEDGMENTS. As always, I am grateful for the support of my work by my wife Janice M. Jerrells and especially her editorial input into this chapter. I thank my friend Garnett Kelsoe, Ph.D., for defining redux for me. The various mentors that have worked with me during my career as a rickettsial researcher are also acknowledged for their invaluable input into my thinking.

REFERENCES

1. Jerrells, T. R., 1988, Mechanisms of immunity to *Rickettsia* species and *Coxiella burnetii*, in: *Biology of Rickettsial Diseases*, Volume II (D. H. Walker, ed.), CRC Press, Boca Raton, pp. 79–100.
2. Kaufmann, S. H. E., 1993, Immunity to intracellular bacteria, in: *Annual Review of Immunology*, Volume 11 (W. E. Paul, C. G. Gathman, and H. Metzgar, eds.), Annual Reviews, Palo Alto, pp. 129–163.
3. Mosmann, T. R., and Coffman, R. L., 1989, Heterogeneity of cytokine secretion patterns and functions of helper T cells, *Adv. Immunol.* **46:**111–147.
4. Mosmann, T. R., Cherwinski, H., Bond, M. W., Giedlin, M. A., and Coffman, R. L., 1986, Two types of murine helper T cell clone. I. Definition according to profiles of lymphokine activities and secreted proteins, *J. Immunol.* **136:**2348–2357.
5. Mosmann, T. R., and Coffman, R. L., 1989, TH1 and TH2 cells: Different patterns of lympho-kine secretion lead to different functional properties, *Annu. Rev. Immunol.* **7:**145–173.
6. Cher, D. J., and Mosmann, T. R., 1987, Two types of murine helper T-cell clone. II. Delayed-type hypersensitivity is mediated by TH1 clones, *J. Immunol.* **138:**3688–3694.
7. Bancroft, G. J., 1993, The role of natural killer cells in innate resistance to infection, *Curr. Opin. Immunol.* **5:**503–510.
8. Hsieh, C.-S., Macatonia, S. E., Tripp, C. S., Wolf, S. F., O'Garra, A., and Murphy, K. M., 1993, Development of TH1 CD4+T cells through IL-12 produced by *Listeria*-infected macrophages, *Science* **260:**547–549.
9. Manetti, R., Parronchi, P., Giudizi, M. G., Piccinni, M.-P., Maggi, E., Trinchieri, G., and Romagnani, S., 1993, Natural killer cell stimulatory factor (interleukin 12 [IL-12]) induces T helper type 1 (Th1)-specific immune responses and inhibits the development of IL-4-producing Th cells, *J. Exp. Med.* **177:**1199–1204.

10. Murphy, E. E., Terres, G., Macatonia, S. E., Hsieh, C.-S., Mattson, J., Lanier, L., Wysocka, M., Trinchieri, G., Murphy, K., and O'Garra, A., 1994, B7 and interleukin 12 cooperate for proliferation and interferon-γ production by mouse T helper clones that are unresponsive to B7 costimulation, *J. Exp. Med.* **180:**223–231.

11. Scott, P., 1993, IL-12: Initiation cytokine for cell-mediated immunity, *Science* **260:**496–497.

12. Trinchieri, G., 1994, Interleukin-12: A cytokine produced by antigen-presenting cells with immunoregulatory functions in the generation of T-helper cells type 1 and cytotoxic lymphocytes, *Blood* **84:**4008–4027.

13. Hendrzak, J. A., and Brunda, M. J., 1995, Interleukin-12: Biologic activity, therapeutic utility, and role in disease, *Lab. Invest.* **72:**619–637.

14. Heinzel, F. P., Sadick, M. D., Holaday, B. J., Coffman, R. L., and Locksley, R. M., 1989, Reciprocal expression of interferon-gamma or interleukin-4 during the resolution or progression of murine leishmaniasis. Evidence for expansion of distinct helper T cell subsets, *J. Exp. Med.* **169:**59–72.

15. Sher, A., Fiorentino, D., Caspar, P., Pearce, E., and Mosmann, T. R., 1991, Production of IL-10 by CD4+ T lymphocytes correlates with down-regulation of Th1 cytokine synthesis in helminth infection, *J. Immunol.* **147:**2713–2716.

16. Groves, M. G., and Osterman, J. V., 1978, Host defenses in experimental scrub typhus: Genetics of natural resistance to infection, *Infect. Immun.* **19:**583–588.

17. Rollwagen, F. M., Dasch, G. A., and Jerrells, T. R., 1986, Mechanisms of immunity to rickettsial infection: Characterization of a cytotoxic effector cell, *J. Immunol.* **136:**1418–1421.

18. Sasaki, T., Mieno, M., Udono, H., Yamaguchi, K., Usui, T., Hara, K., Shiku, H., and Nakayama, E., 1990, Roles of CD4+ and CD8+ cells, and the effect of administration of recombinant murine interferon γ in listerial infection, *J. Exp. Med.* **171:**1141–1154.

19. Hess, J., Ladel, C., Miko, D., and Kaufmann, S. H. E., 1996. *Salmonella typhimurium* aroA⁻ infection in gene-targeted immunodeficient mice: major role of CD4+ TCR-alpha beta cells and IFN-gamma in bacterial clearance independent of intracellular location, *J. Immunol.* **156:**3321–3326.

20. Ohga, S., Yoshikai, Y., Takeda, Y., Hiromatsu, K., and Nomoto, K., 1990, Sequential appearance of γ/δ- and α/β-bearing T cells in the peritoneal cavity during an intraperitoneal infection with *Listeria monocytogenes, Eur. J. Immunol.* **20:**533–538.

21. Emoto, M., Naito, T., Nakamura, R., and Yoshikai, Y., 1993, Different appearance of γδ T cells during salmonellosis between *Ity^r* and *Ity^s* mice, *J. Immunol.* **150:**3411–3420.

22. Shirai, A., Catanzaro, P. J., Phillips, S. M., and Osterman, J. V., 1976, Host defenses in experimental scrub typhus: Role of cellular immunity in heterologous protection, *Infect. Immun.* **14:**39–46.

23. Catanzaro, P. J., Shirai, A., Agniel, L. D., Jr., and Osterman, J. V., 1977, Host defenses in experimental scrub typhus: Role of spleen and peritoneal exudate lymphocytes in cellular immunity, *Infect. Immun.* **18:**118–123.

24. Jerrells, T. R., Palmer, B. A., and MacMillan, J. G., 1984, Cellular mechanisms of innate and acquired immunity to *Rickettsia tsutsugamushi*, in: *Microbiology—1984* (L. Leive and D. Schlessinger, eds.), American Society for Microbiology, Washington, DC, pp. 277–281.

25. Crist, A. E., Jr., Wisseman, C. L., Jr., and Murphy, J. R., 1984, Characteristics of lymphoid cells that adoptively transfer immunity to *Rickettsia mooseri* infection in mice, *Infect. Immun.* **44:**55–60.

26. Jerrells, T. R., and Osterman, J. V., 1982, Host defenses in experimental scrub typhus: Delayed-type hypersensitivity responses of inbred mice, *Infect. Immun.* **35:**117–123.

27. Jerrells, T. R., and Eisemann, C. S., 1983, Role of T-lymphocytes in production of antibody to antigens of *Rickettsia tsutsugamushi* and other *Rickettsia* species, *Infect. Immun.* **41:**666–674.

28. Jerrells, T. R., Mallavia, L. P., and Hinrichs, D. J., 1975, Detection of long-term cellular immunity to *Coxiella burneti* as assayed by lymphocyte transformation, *Infect. Immun.* **11:**280–286.

29. Coonrod, J. D., and Shepard, C. C., 1971, Lymphocyte transformation in rickettsioses, *J. Immunol.* **106:**209–216.

30. Bourgeois, A. L., Dasch, G. A., and Strong, D. M., 1980, In vitro stimulation of human peripheral blood lymphocytes by soluble and membrane fractions of renografin-purified typhus group rickettsiae, *Infect. Immun.* **27:**483–491.

31. Kenyon, R. H., Ascher, M. S., Kishimoto, R. A., and Pedersen, C. E., Jr., 1977, In vitro guinea pig leukocyte reactions to *Rickettsia rickettsii*, *Infect. Immun.* **18:**840–846.

32. Jerrells, T. R., and Osterman, J. V., 1983, Development of specific and cross-reactive lymphocyte proliferative responses during chronic immunizing infections with *Rickettsia tsutsugamushi*, *Infect. Immun.* **40:**147–156.

33. MacMillan, J. G., Rice, R. M., and Jerrells, T. R., 1985, Development of antigen-specific cell-mediated immune responses after infection of cynomolgus monkeys (*Macaca fascicularis*) with *Rickettsia tsutsugamushi*, *J. Infect. Dis.* **152:**739–749.

34. Jerrells, T. R., Jarboe, D. L., and Eisemann, C. S., 1986, Cross-reactive lymphocyte responses and protective immunity against other spotted fever group rickettsiae in mice immunized with *Rickettsia conorii*, *Infect. Immun.* **51:**832–837.

35. Jarboe, D. L., Eisemann, C. S., and Jerrells, T. R., 1986, Production and characterization of cloned T-cell hybridomas that are responsive to *Rickettsia conorii* antigens, *Infect. Immun.* **52:**326–330.

36. Williams, N. M., Granstrom, D. E., and Timoney, P. J., 1994, Humoral antibody and lymphocyte blastogenesis responses in BALB/c, C3H/HeJ, and AKR/N mice following *Ehrlichia risticii* infection, *Res. Vet. Sci.* **56:**284–289.

37. Palmer, B. A., Hetrick, F. M., and Jerrells, T. R., 1984, Production of gamma interferon in mice immune to *Rickettsia tsutsugamushi*, *Infect. Immun.* **43:**59–65.

38. Palmer, B. A., Hetrick, F. M., and Jerrells, T. R., 1984, Gamma interferon production in response to homologous and heterologous strain antigens in mice chronically infected with *Rickettsia tsutsugamushi*, *Infect. Immun.* **46:**237–244.

39. Hinrichs, D. J., and Jerrells, T. R., 1976, In vitro evaluation of immunity to *Coxiella burnetii*, *J. Immunol.* **117:**996–1003.

40. Nacy, C. A., and Osterman, J. V., 1979, Host defenses in experimental scrub typhus: Role of normal and activated macrophages, *Infect. Immun.* **26:**744–750.

41. Nacy, C. A., Leonard, E. J., and Meltzer, M. S., 1981, Macrophages in resistance to rickettsial infections: Characterization of lymphokines that induce rickettsiacidal activity in macrophages, *J. Immunol.* **126:**204–207.

42. Gage, K. L., and Jerrells, T. R., 1992, Demonstration and partial characterization of antigens of *Rickettsia rhipicephali* that induce cross-reactive cellular and humoral immune responses to *Rickettsia rickettsii*, *Infect. Immun.* **60:**5099–5106.

43. Murata, M., and Kawamura, A., Jr., 1977, Restoration of the infectivity of *Rickettsia tsutsugamushi* to susceptible animals by passage in athymic nude mice, *Jpn. J. Exp. Med.* **47:**385–391.

44. Kenyon, R. H., and Pedersen, C. E., Jr., 1980, Immune responses to *Rickettsia akari* infection in congenitally athymic nude mice, *Infect. Immun.* **28:**310–313.

45. Kokorin, I. N., Kabanova, E. A., Shirokova, E. M., Abrosimova, G. E., Rybkina, N. N., and Pushkareva, V. I., 1982, Role of T lymphocytes in *Rickettsia conorii* infection, *Acta Virol.* **26:**91–97.

46. Messick, J. B., and Rikihisa, Y., 1994, Inhibition of binding, entry, or intracellular proliferation of *Ehrlichia risticii* in P388D1 cells by anti-*E. risticii* serum, immunoglobulin G, or Fab fragment, *Infect. Immun.* **62:**3156–3161.

47. Michetti, P., Mahan, M. J., Slauch, J. M., Mekalanos, J. J., and Neutra, M. R., 1992, Monoclonal secretory immunoglobulin A protects mice against oral challenge with the invasive pathogen *Salmonella typhimurium*, *Infect. Immun.* **60:**1786–1792.

48. Barnewall, R. E., and Rikihisa, Y., 1994, Abrogation of gamma interferon-induced inhibition of *Ehrlichia chaffeensis* infection in human monocytes with iron-transferrin, *Infect. Immun.* **62:**4804–4810.

49. Turco, J., and Winkler, H. H., 1984, Effect of mouse lymphokines and cloned mouse interferon-γ on the interaction of *Rickettsia prowazekii* with mouse macrophage-like RAW264.7 cells, *Infect. Immun.* **45:**303–308.

50. Chan, J., Xing, Y., Magliozzo, R. S., and Bloom, B. R., 1992, Killing of virulent *Mycobacterium tuberculosis* by reactive nitrogen intermediates produced by activated murine macrophages, *J. Exp. Med.* **175:**1111–1122.

51. Gebran, S. J., Yamamoto, Y., Newton, C., Klein, T. W., and Friedman, H., 1994, Inhibition of *Legionella pneumophila* growth by gamma interferon in permissive A/J mouse macrophages: Role of reactive oxygen species, nitric oxide, tryptophan, and iron(III), *Infect. Immun.* **62:**3197–3205.

52. Feng, H. M., and Walker, D. H., 1993, Interferon-gamma and tumor necrosis factor-alpha exert their antirickettsial effect via induction of synthesis of nitric oxide, *Am. J. Pathol.* **143:**1016–1023.

53. Jerrells, T. R., Turco, J., Winkler, H. H., and Spitalny, G. L., 1986, Neutralization of lympho-kine-mediated antirickettsial activity of fibroblasts and macrophages with monoclonal antibody specific for murine interferon gamma, *Infect. Immun.* **51:**355–359.

54. Turco, J., and Winkler, H. H., 1983, Inhibition of the growth of *Rickettsia prowazekii* in cultured fibroblasts by lymphokines, *J. Exp. Med.* **157:**974–986.

55. Wisseman, C. L., Jr., and Waddell, A., 1983, Interferon-like factors from antigen- and mitogen-stimulated human leukocytes with antirickettsial and cytolytic actions on *Rickettsia prowazekii*. Infected human endothelial cells, fibroblasts, and macrophages, *J. Exp. Med.* **157:**1780–1793.

56. Turco, J., and Winkler, H. H., 1983, Comparison of the properties of antirickettsial activity and interferon in mouse lymphokines, *Infect. Immun.* **42:**27–32.

57. Turco, J., Thompson, H. A., and Winkler, H. H., 1984, Interferon-γ inhibits the growth of *Coxiella burnetii* in mouse fibroblasts, *Infect. Immun.* **45:**781–783.

58. Szalay, G., Hess, J., and Kaufmann, S. H. E., 1995, Restricted replication of *Listeria monocytogenes* in a gamma interferon-activated murine hepatocyte line, *Infect. Immun.* **63:**3187–3195.

59. Jerrells, T. R., and Geng, P., 1994, The role of tumor necrosis factor in host defense against scrub typhus rickettsiae. II. Differential induction of tumor necrosis factor-alpha production by *Rickettsia tsutsugamushi* and *Rickettsia conorii*, *Microbiol. Immunol.* **38:**713–719.

60. Manor, E., and Sarov, I., 1990, Tumor necrosis factor alpha and prostaglandin E_2 production by human monocyte-derived macrophages infected with spotted fever group rickettsiae, *Ann. N.Y. Acad. Sci.* **590:**157–167.

61. Geng, P., and Jerrells, T. R., 1994, The role of tumor necrosis factor in host defense against scrub typhus rickettsiae. I. Inhibition of growth of *Rickettsia tsutsugamushi*, Karp strain, in cultured murine embryonic cells and macrophages by recombinant tumor necrosis factor-alpha, *Microbiol. Immunol.* **38:**703–711.

62. Sporn, L. A., and Marder, V. J., 1996, Interleukin-1α production during *Rickettsia rickettsii* infection of cultured endothelial cells: Potential role in autocrine cell stimulation, *Infect. Immun.* **64:**1609–1613.

63. Oster, C. N., Kenyon, R. H., and Pedersen, C. E., Jr., 1978, Suppression of cellular immune responses in guinea pigs infected with spotted fever group rickettsiae, *Infect. Immun.* **22:**411–417.

64. Jerrells, T. R., 1985, Immunosuppression associated with the development of chronic infections with *Rickettsia tsutsugamushi*: Adherent suppressor cell activity and macrophage activation, *Infect. Immun.* **50:**175–182.

65. Jerrells, T. R., and Hickman, C. J., 1987, Down regulation of Ia antigen expression on inflamma-tory macrophages by factors produced during acute infections with *R. tsutsugamushi*, in: *Immune Regulation by Characterized Polypeptides*, Volume 41, UCLA Symposia on Molecular and Cellular Biology (G. Goldstein, J. F. Bach, and H. Wigzell, eds.), Liss, New York, pp. 659–668.

66. Koster, F. T., Williams, J. C., and Goodwin, J. S., 1985, Cellular immunity in Q fever: Specific lymphocyte unresponsiveness in Q fever endocarditis, *J. Infect. Dis.* **152:**1283–1289.

67. Koster, F. T., Williams, J. C., and Goodwin, J. S., 1985, Cellular immunity in Q fever: Modulation of responsiveness by a suppressor T cell–monocyte circuit, *J. Immunol.* **135:**1067–1072.

68. Damrow, T. A., Williams, J. C., and Waag, D. M., 1985, Suppression of in vitro lymphocyte proliferation in C57BL/10 ScN mice vaccinated with phase I *Coxiella burnetii, Infect. Immun.* **47:**149–156.

69. Heinzen, R. A., Hayes, S. F., Peacock, M. G., and Hackstadt, T., 1993, Directional actin polymerization associated with spotted fever group Rickettsia infection of Vero cells, *Infect. Immun.* **61:**1926–1935.

3

Cytokines Influencing Infections by *Rickettsia* Species

JENIFER TURCO and HERBERT H. WINKLER

1. INTRODUCTION

The genus *Rickettsia* is a group of obligate intracellular bacteria that grow (both in nature and in the laboratory) only within vertebrate or arthropod cells. Unlike many other intracellular bacteria, intracellular rickettsiae are not separated from the cytoplasm of their host cells by the membranes of vacuoles (phagosomes or phagolysosomes). Rather, rickettsiae grow within the cytoplasm of their host cells with host cell cytoplasm immediately outside their own surface layers.

Members of the genus have traditionally been divided into three groups: (1) the typhus group [including *Rickettsia prowazekii* and *Rickettsia typhi*, the causative agents of epidemic (louse-borne) typhus and endemic typhus, respectively]; (2) the spotted fever group (including *Rickettsia rickettsii*, the etiologic agent of Rocky Mountain spotted fever; *Rickettsia conorii*, the etiologic agent of Mediterranean spotted fever or Boutonneuse fever; *Rickettsia akari*, the causative agent of rickettsialpox; and *Rickettsia australis*, which causes Queensland tick typhus); and (3) the scrub typhus group (including *Rickettsia tsutsugamushi*, the etiologic agent of scrub typhus).[1] However, because *R. tsutsugamushi* exhibits many phenotypic and ge-

JENIFER TURCO • Department of Biology, Valdosta State University, Valdosta, Georgia 31698-0015.   HERBERT H. WINKLER • Laboratory of Molecular Biology, Department of Microbiology and Immunology, University of South Alabama College of Medicine, Mobile, Alabama 36688.

Rickettsial Infection and Immunity, edited by Anderson *et al.* Plenum Press, New York, 1997.

notypic differences from other members of the genus *Rickettsia,* classification of this bacterium in another genus (as *Orientia tsutsugamushi*) has been proposed.[2] In addition, sequence analysis of the rickettsial ribosomal RNA genes has provided evidence that *R. akari, R. australis,* and the ELB agent (*Rickettsia felis*[3]) belong to a unique rickettsial group.[4]

This review discusses the influence of interferons and tumor necrosis factors on rickettsial infections. Cytokines modify the interactions between rickettsiae and their host cells in several ways: inhibition of rickettsial growth occurs in cytokine-treated host cells; killing of intracellular rickettsiae occurs in cytokine-treated host cells; the cytokines and the rickettsiae in combination have cytotoxic effects on the host cells; and the ability of rickettsiae to initially infect cytokine-treated host cells is suppressed (Fig. 1; reviewed in Refs. 5–9). In the studies of rickettsial infections in various hosts (cultured cells or animals) described herein, the cytokines used are of the same species as the host, unless otherwise noted. Studies with typhus, spotted fever, and scrub typhus rickettsiae are discussed; and emphasis is given to information (including primary data) not covered in earlier reviews. In particular, evidence supporting the importance of both nitric oxide synthase-dependent and -independent antirickettsial mechanisms in cytokine-treated host cells is summarized and reviewed (Fig. 1).

2. STUDIES INVOLVING INTERFERONS

2.1. Induction of Interferons by Rickettsiae

Alpha/beta interferon (IFN-α/β or type I interferon) and gamma interferon (IFN-γ) are induced by all species of *Rickettsia* tested. *R. prowazekii,*[10] *R. typhi,*[11] and *R. tsutsugamushi*[12] are able to induce the production of IFN-α/β in cultured cells. IFN-α/β is also found in the sera of mice inoculated with *R. prowazekii*[13]; and both IFN-α/β and IFN-γ are detected in the sera of mice inoculated with *R. conorii* Malish 7.[14–16] Naive cynomolgus monkeys produce IFN-γ after challenge with *R. tsutsugamushi* Karp; and monkeys previously infected with *R. tsutsugamushi* Karp produce IFN-γ after they are rechallenged.[17] Cultured peripheral blood leukocytes from either naive or previously infected cynomolgus monkeys also produce IFN-γ on addition of *R. tsutsugamushi* Karp antigen.[17] Furthermore, mice immunized with *R. tsutsugamushi* Gilliam produce IFN-γ after rechallenge; and cultured spleen cells and lymph node cells from immunized mice produce IFN-γ after they are exposed to *R. tsutsugamushi* antigen *in vitro.*[18,19] Finally, human peripheral blood leukocytes obtained from individuals who have recovered from infection with *R. prowazekii* or *R. typhi* produce IFN-γ after exposure to *R. prowazekii* antigen *in vitro.*[20]

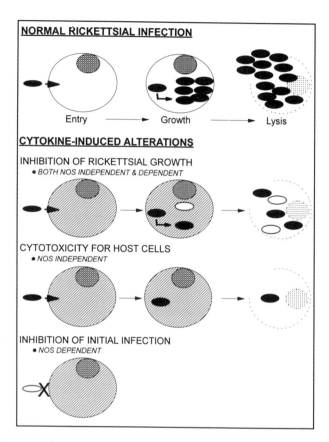

FIGURE 1. Overview of cytokine-induced changes in rickettsia–host cell interactions. In the usual infection of cultured cells (top panel), *R. prowazekii* organisms enter their host cells by induced phagocytosis, and grow to fill the cytoplasm. Lysis of the host cells then occurs. In cytokine-treated host cells, several changes are observed in the rickettsial infection. First, *R. prowazekii* organisms may enter the host cell and multiply at a reduced rate; and some of the intracellular rickettsiae may be killed. Such restriction of rickettsial growth in cytokine-treated host cells involves both nitric oxide synthase (NOS)-independent and NOS-dependent mechanisms. Second, the combination of *R. prowazekii* infection and cytokine treatment may have a cytotoxic effect on the host cells. In some situations the cytotoxic effect occurs within hours after infection and does not require growth of the rickettsiae; in other situations, some growth of the rickettsiae may occur before the host cells die. These cytotoxic effects on the host cells are NOS-independent. Third, when cultured cells that have been pretreated with cytokines are incubated with *R. prowazekii* organisms, many of the extracellular rickettsiae are killed by a NOS-dependent mechanism. The rickettsiae are therefore prevented form infecting the cells.

TABLE I
Inhibition of Growth of *R. prowazekii* in X-Irradiated Human MRC-5
Fibroblasts by IFN-β_{ser} and IFN-γ

Rickettsial strain	Cell treatment[a]	No. of expts.	Infection on day 2 (% of value on day 0)[b]			Rickettsial growth (% control)[c]
			% R	RI	NR	
Madrid E	Control	3	105 ± 1	978 ± 89	1027 ± 95	100
	IFN-β_{ser} (1000 U/ml)	3	94 ± 2	610 ± 108	583 ± 118	55 ± 6**
	IFN-γ (100 U/ml)	3	73 ± 7	229 ± 44	175 ± 49	16 ± 3***††
Breinl	Control	2	106 ± 11	1310 ± 537	1512 ± 721	100
	IFN-β_{ser} (1000 U/ml)	2	106 ± 21	579 ± 218	700 ± 349	45 ± 2**
	IFN-γ (100 U/ml)	2	63 ± 2	168 ± 3	106 ± 6	12 ± 6*†

[a]Untreated or interferon-treated MRC-5 cells were infected with rickettsiae on day 0. Cells were treated with interferon both before and after infection.
[b]The values for % R (percentage of cells infected), RI (rickettsiae per infected cell), and NR (rickettsiae per cell) on day 2 are expressed as percentages of the respective values on day 0. Each value represents the mean ± SEM. On day 0, the values of % R and RI in the Madrid E-infected cultures were 88 ± 2% and 7.3 ± 0.7, 95 ± 0% and 8.5 ± 1.3, and 87 ± 1% and 6.2 ± 0.7 in the untreated, IFN-β_{ser}-treated, and IFN-γ-treated cultures, respectively. The values of % R and RI in the Breinl-infected cultures on day 0 were 73 ± 11% and 5.5 ± 2.6, 69 ± 14% and 5.0 ± 2.1, and 71 ± 1% and 2.9 ± 0.6 in the untreated, IFN-β_{ser}-treated, and IFN-γ-treated cultures, respectively.
[c]The values of NR in the treated cultures on day 2 (expressed as percentages of the values of NR on day 0) are expressed as percentages of the respective control values. Each value represents the mean ± SEM. Values that were significantly different from 100 (as determined by the one-sample t test) are indicated as follows: *$p < 0.1$, **$p < 0.05$, or ***$p < 0.01$. A significant difference between IFN-γ-treated and IFN-β_{ser}-treated cultures infected with a given rickettsial strain (as determined by the two-sample t test) is indicated as follows: †$p < 0.1$, or ††$p < 0.01$.

2.2. Effects of Interferons on Rickettsial Growth

Alpha / Beta Interferon

Kazar *et al.*[21] first demonstrated the ability of IFN-α/β to inhibit the growth of *R. akari* in mouse L929 cells. Not only does the addition of IFN-α/β to L929 cells infected with *R. prowazekii* Madrid E suppress rickettsial growth, but also the endogenously produced IFN-α/β in untreated, strain Madrid E-infected L929 cell cultures limits the growth of the rickettsiae.[10,22] In contrast, *R. prowazekii* Breinl (a virulent strain) is resistant to IFN-α/β in mouse L929 cells compared with *R. prowazekii* Madrid E (an avirulent strain).[23] However, in human MRC-5 fibroblasts, IFN-β_{ser} similarly inhibits the growth of both the Madrid E and Breinl strains of *R. prowazekii* (Table I). (In IFN-β_{ser}, serine is substituted for cysteine at amino acid position 17.) Hanson[24] determined that the multiplication of *R. rickettsii* is slightly inhibited by IFN-α/β in the mouse fibroblast lines BALB/3T3 clone A31 and C3H/10T1/2 clone 8. In addition, she evaluated *R. tsutsugamushi* strains Karp, Gilliam, and TA716 for sensitivity to IFN-α/β and noted a correlation (confounded by the weak inhibitions observed) between ability to cause lethal infections in mice and a lack of sensitivity to IFN-α/β in mouse fibroblast lines. Growth of the Karp strain (which causes lethal infections in both BALB/c and C3H/He mice[25]) is not significantly inhibited by IFN-α/β in either BALB/3T3 or C3H/10T1/2 cells.[24] In contrast, the growth of strain TA716 (which does not cause lethal infections in these mice[25]) is slightly (but variably) inhibited by IFN-α/β in both BALB/3T3 and C3H/10T1/2 cells.[24] Finally, growth of the Gilliam strain (which causes lethal infections in C3H/He mice but not in BALB/c mice[25]) is not significantly inhibited by IFN-α/β in C3H/10T1/2 cells, but is slightly (but variably) inhibited by IFN-α/β in BALB/3T3 cells.[24] Although these *in vitro* studies indicate that IFN-α/β may have a modest antirickettsial effect, to our knowledge, convincing *in vivo* evidence is not available to indicate that IFN-α/β plays a major role in antirickettsial host defense. The possibility that IFN-α/β is an important antirickettsial host defense has not been eliminated, however, as the effects of neutralization of IFN-α/β in rickettsia-infected or -challenged animals have not been evaluated.

Gamma Interferon

IFN-γ is able to dramatically inhibit the growth of *R. prowazekii* Breinl and *R. prowazekii* Madrid E in cultured mouse and human cells.[8,20,23,26,27] The inhibition of either strain of *R. prowazekii* by IFN-γ is significantly greater than the inhibition by IFN-β_{ser} in MRC-5 fibroblasts (Table I); and in L929 cells IFN-γ treatment causes significantly greater inhibition of *R. prowazekii* Madrid E than does IFN-α/β.[23] The replication of *R. rickettsii* is much more inhibited by rat IFN-γ than murine IFN-α/β in the mouse fibroblast lines BALB/3T3 clone A31 and C3H/10T1/2 clone 8.[24] Studies with *R. conorii* indicate that IFN-γ restricts the growth of: (1) the Casablanca strain in mouse macrophages,[28] (2) the Moroccan strain in mouse macrophagelike J774A.1 cells,[29] and (3) the Malish 7 strain in mouse L929 cells.[30] However, pretreatment of human HEp-2 cells with IFN-γ does not inhibit the growth of *R. conorii* Casablanca within these cells.[31] In addition, convincing *in vivo* evidence demonstrating the importance of IFN-γ as an antirickettsial host defense in mice was obtained in studies with *R. conorii* Malish 7.[14,15] IFN-γ restricts the replication of *R. tsutsugamushi* Karp in macrophages obtained from C57BL/6mice.[32] However, the Karp and TA716 strains of *R. tsutsugamushi* are resistant to rat IFN-γ in the BALB/3T3 clone A31 and C3H/10T1/2 clone 8 fibroblasts, whereas the Gilliam strain is sensitive to rat IFN-γ in the first and resistant to IFN-γ in the second of these cell lines.[24]

Discussion and Perspective

Experimental evidence thus indicates that it is difficult (if not impossible) to make generalizations concerning the sensitivity (or insensitivity) of rickettsiae to various IFNs. For example, *R. prowazekii* strains differ in sensitivity to IFN-α/β in mouse L929 cells.[23]; *R. tsutsugamushi* strains differ in sensitivity to IFN-α/β in mouse BALB/3T3 cells and/or C3H/10T1/2 cells[24]; and *R. tsutsugamushi* strains differ in sensitivity to rat IFN-γ in mouse BALB/3T3 cells.[24] In addition, sensitivity of a given rickettsial strain [e.g., *R. tsutsugamushi* Gilliam,[24] *R. prowazekii* Madrid E (Table II), and *R. prowazekii* Breinl (Table II)] to IFN-γ is influenced by the cell line used. Finally, although the inhibitory effect of IFN-γ on a particular rickettsial strain in a particular cell line is generally more dramatic than that of IFN-α/β [consider, for example, *R. prowazekii* Madrid E in mouse L929 cells,[10,23] *R. prowazekii* Madrid E or Breinl in human MRC-5 fibroblasts (Table I), *R. rickettsii* in BALB/3T3 and C3H/10T1/2 cells,[24] and *R. tsutsugamushi* Gilliam in BALB/3T3 cells[24]], the opposite is true for *R. tsutsugamushi* TA716 (which is insensitive to rat IFN-γ but slightly sensitive to IFN-α/β in BALB/3T3 cells and C3H/10T1/2 cells[24]).

Studies Concerning the Mechanisms by Which Interferons Inhibit Growth of Rickettsiae

The effects of IFN-γ on growth of the Madrid E and Breinl strains of *R. prowazekii* were examined in additional human cell lines to obtain information about which established, IFN-γ-induced activities correlated with inhibition of rickettsial growth (Table II). Multiplication of *R. prowazekii* Madrid E was markedly inhibited by IFN-γ in human GM2767A cells, even though these cells do not exhibit an increase in the activity of the double-stranded RNA-activated protein kinase in response to IFN-γ.[33] Thus, an increase in the activity of this enzyme was unlikely to be required for marked inhibition of rickettsial growth to occur in GM2767A cells. In contrast, growth of the Madrid E and Breinl strains of *R. prowazekii* was only modestly suppressed in IFN-γ-treated RD114 cells and IFN-γ-treated RD114-A2 cells. Because, in response to IFN-γ, RD114 cells restrict retrovirus replication and RD114-A2 cells restrict retrovirus and encephalomyocarditis virus replication,[34] the mechanisms responsible for inhibition of these viruses in these cells were apparently not sufficient for marked inhibition of rickettsial growth to occur.

Studies aimed at defining the mechanisms by which IFN-γ restricts the growth of *R. prowazekii* strains in mouse and human cells have suggested that multiple mechanisms of inhibition are involved and that the Madrid E and Breinl strains differ in sensitivity to particular IFN-induced inhibitory mechanisms. The involvement of multiple mechanisms was first suggested by experiments that were designed to evaluate the possible role of amino acid deprivation in limiting the growth of *R. prowazekii* Madrid E in mouse L929 cells and human fibroblasts.[26] Reconstitution of the depleted tryptophan pool in IFN-γ-treated human fibroblasts does not relieve the inhibition of the growth of *R. prowazekii* Madrid E in IFN-γ-treated human fibroblasts,[26] although tryptophan supplementation does

TABLE II
Effect of IFN-γ on Growth of *R. prowazekii* Madrid E and Breinl in Several X-Irradiated Human Cell Lines

Cell line	Treatment with IFN-γ relative to time of infection[a]	No of expts.	Madrid E-infected cultures				No of expts.	Breinl-infected cultures			
			Infection on day 2 (% of value on day 0)[b]			Rickettsial growth (% control)[c]		Infection on day 2 (% of value on day 0)[b]			Rickettsial growth (% control)[c]
			% R	RI	NR			% R	RI	NR	
MRC-5	None	5	106 ± 2	1167 ± 141	1247 ± 170	100	4	96 ± 8	1302 ± 270	1306 ± 375	100
	Before	2	72 ± 5	451 ± 71	317 ± 30	22 ± 6*	2	72 ± 5	129 ± 10	92 ± 1	8 ± 1***
	After	2	99 ± 0	547 ± 55	541 ± 54	35 ± 2**	2	74 ± 7	149 ± 17	108 ± 2	10 ± 1***†
	Both	5	78 ± 5	230 ± 29	185 ± 34	15 ± 2****	4	55 ± 7	127 ± 21	73 ± 17	8 ± 4****
GM2767A	None	4	112 ± 9	1209 ± 277	1452 ± 435	100			Not tested		
	Before	4	74 ± 8	208 ± 65	162 ± 60	10 ± 1****					
HEp-2	None	4	98 ± 6	2300 ± 226	2255 ± 260	100	4	100 ± 5	1696 ± 157	1638 ± 195	100
	Both	4	85 ± 6	1478 ± 229	1240 ± 132	55 ± 2****	4	73 ± 8	241 ± 39	187 ± 49	11 ± 2*****†††
BT-20	None	2	109 ± 8	1344 ± 280	1537 ± 374	100	2	83 ± 8	1416 ± 189	1142 ± 168	100
	Both	2	113 ± 25	1065 ± 308	1014 ± 341	67 ± 10*	2	61 ± 9	181 ± 18	111 ± 21	11 ± 2*****††
RD114	None	5	115 ± 6	1614 ± 259	1918 ± 374	100	9	119 ± 14	1866 ± 297	2528 ± 794	100
	Both	5	96 ± 5	1191 ± 245	1174 ± 302	60 ± 5***	9	113 ± 12	1245 ± 214	1614 ± 501	66 ± 6****
RD114-A2	None	4	129 ± 17	1827 ± 166	2435 ± 548	100	3	122 ± 16	1597 ± 210	2003 ± 437	100
	Both	4	115 ± 15	1519 ± 222	1862 ± 510	75 ± 10	3	114 ± 9	1285 ± 263	1524 ± 379	74 ± 5**

[a] Cells were infected with rickettsiae on day 0. Cultures were treated with IFN-γ (200 U/ml for GM2767A, MRC-5, HEp-2, and BT-20; and 1000 U/ml for RD114 and RD114-A2) before infection only, after infection only, or both before and after infection.

[b] The values for % R (percentage of cells infected), RI (rickettsiae per infected cell), and NR (rickettsiae per cell) on day 2 are expressed as percentages of the respective values on day 0. Each value represents the mean ± SEM. On day 0, the average values for % R and RI in the various cultures ranged from 54 to 89% and from 2.0 to 7.5, respectively.

[c] The values of NR in the treated cultures on day 2 (expressed as percentages of the values of NR on day 0) are expressed as percentages of the respective control values. Each value represents the mean ± SEM. Values that were significantly different from 100 (as determined by the one-sample t test) are indicated as follows: *$p < 0.1$, **$p < 0.05$, ***$p < 0.01$, or ****$p < 0.001$. A significant difference between treated cultures infected with the Breinl strain and the corresponding cultures infected with the Madrid E strain (as determined by the two-sample t test) is indicated by †$p < 0.025$, ††$p < 0.005$, or †††$p < 0.001$.

alleviate the IFN-γ-induced inhibition of the growth of certain other intracellular microorganisms in human fibroblasts.[35] In the human cell lines HEp-2 and BT-20, the Breinl strain was significantly more sensitive to IFN-γ than the Madrid E strain (Table II). Indoleamine 2,3-dioxygenase (tryptophan degrading) activity did not correlate with the observed inhibition, as this activity was detected in IFN-γ-treated HEp-2 cells but not in IFN-γ-treated BT-20 cells (J. Turco and H. H. Winkler, unpublished data). The results with IFN-γ-treated BT-20 cells indicate that the Breinl strain must be very sensitive (and the Madrid E strain much less sensitive) to an inhibitory mechanism other than tryptophan degradation in these cells. IFN-γ also does not induce detectable indoleamine 2,3-dioxygenase activity in mouse L929 cells yet it markedly suppresses the growth of both *R. prowazekii* Madrid E and Breinl in these cells.[23,26] These findings indicate the importance of mechanisms other than tryptophan deprivation in IFN-γ-induced suppression of growth of *R. prowazekii* in both human and mouse cells.

Studies with *R. prowazekii* and IFN-γ have also provided evidence that there is an IFN-γ-induced, nitric oxide synthase-independent, inhibitory mechanism that is common to mouse L929 cells and human fibroblasts. This nitric oxide synthase-independent mechanism plays a major role in suppressing the growth of *R. prowazekii* Madrid E and Breinl in IFN-γ-treated L929 cells, as inhibition of nitric oxide synthase does not relieve the IFN-γ-induced suppression of the growth of these strains.[36] That this mechanism is also active in IFN-γ-treated human fibroblasts was suggested by the fact that Madrid E-derived, IFN-resistant *R. prowazekii* organisms are resistant to IFN-γ both in mouse L929 cells[23] and in human fibroblasts (Tables III and IV).

It is likely that *R. conorii* is insensitive to one or more IFN-γ-induced growth inhibitory mechanisms that are able to suppress the growth of *R. prowazekii*. For example, Manor and Sarov[31] found that pretreatment of HEp-2 cells with IFN-γ alone does not inhibit the growth of *R. conorii* Casablanca in these human cells. In contrast, we found that IFN-γ markedly restricted the growth of *R. prowazekii* Breinl in HEp-2 cells (Table II). In addition, we found that IFN-γ-induced inhibition of the growth of *R. prowazekii* Madrid E and Breinl in L929 cells is not relieved by inhibition of nitric oxide synthase.[36] However, Feng and Walker[30] observed that inhibition of the growth of *R. conorii* Malish 7 in L929 cells by IFN-γ is alleviated by inhibition of nitric oxide synthase; and Keysary *et al.*[29] found that inhibition of nitric oxide synthase relieves the IFN-γ-induced inhibition of growth of the Moroccan strain of *R. conorii* in mouse macrophagelike J774A.1 cells. Nevertheless, the hypothesis that *R. conorii* is sensitive to nitric oxide synthase-independent as well as nitric oxide synthase-dependent inhibitory mechanisms remains tenable, as the nitric oxide synthase inhibitor did not completely relieve the suppression of growth of *R. conorii* in cytokine-treated L929 cell cultures.[30] These findings illustrate the pitfall of generalizing to all species of rickettsiae from experiments done with one species.

TABLE III
Resistance of *R. prowazekii* Strains 83-2P and 103-2P to IFN-γ in X-Irradiated Human MRC-5 Fibroblasts

Rickettsial strain	Cell treatment[a]	No. of expts.	Cultures treated before infection				Cultures treated after infection			
			Infection on day 2 (% of value on day 0)[b]			Rickettsial growth (% control)[c]	Infection on day 2 (% of value on day 0)[b]			Rickettsial growth (% control)[c]
			% R	RI	NR		% R	RI	NR	
Breinl	Control	2	90 ± 2	731 ± 39	657 ± 51	100	88 ± 2	761 ± 72	673 ± 77	100
	IFN-γ	2	75 ± 11	97 ± 15	76 ± 22	11 ± 2**	80 ± 7	106 ± 3	85 ± 10	13 ± 0***†††
	Trp	2	99 ± 1	671 ± 119	661 ± 114	99 ± 10	83 ± 3	629 ± 67	520 ± 37	78 ± 3
	Trp + IFN-γ	2	73 ± 8	108 ± 11	81 ± 17	12 ± 2**	79 ± 8	110 ± 7	87 ± 15	13 ± 1***
Madrid E	Control	4	98 ± 6	736 ± 158	709 ± 137	100	101 ± 6	906 ± 169	903 ± 158	100
	IFN-γ	4	109 ± 26	277 ± 66	285 ± 71	43 ± 13**	100 ± 8	441 ± 104	429 ± 92	46 ± 3****
	Trp	4	105 ± 11	781 ± 108	837 ± 159	127 ± 30	102 ± 7	881 ± 168	887 ± 152	98 ± 1
	Trp + IFN-γ	4	122 ± 34	278 ± 45	333 ± 96	53 ± 19	97 ± 10	418 ± 94	395 ± 75	43 ± 2****
83-2P	Control	3	102 ± 8	648 ± 148	696 ± 216	100	95 ± 9	524 ± 129	530 ± 173	100
	IFN-γ	3	89 ± 4	304 ± 18	274 ± 26	47 ± 8**	107 ± 3	568 ± 50	610 ± 72	144 ± 36††
	Trp	3	94 ± 2	622 ± 141	578 ± 118	91 ± 8	101 ± 6	713 ± 207	753 ± 263	143 ± 15
	Trp + IFN-γ	3	105 ± 9	335 ± 14	355 ± 44	60 ± 11	105 ± 5	542 ± 99	582 ± 136	127 ± 29
103-2P	Control	4	89 ± 6	717 ± 106	623 ± 84	100	94 ± 2	684 ± 134	637 ± 118	100
	IFN-γ	4	90 ± 8	549 ± 72	475 ± 47	78 ± 4***†	92 ± 4	672 ± 124	615 ± 120	101 ± 16††
	Trp	4	90 ± 5	502 ± 49	455 ± 51	76 ± 9*	90 ± 4	738 ± 134	656 ± 113	105 ± 10
	Trp + IFN-γ	4	92 ± 2	396 ± 37	362 ± 33	63 ± 13*	91 ± 4	673 ± 119	601 ± 87	101 ± 13

[a]MRC-5 cells were infected with rickettsiae on day 0. In cultures treated before infection, IFN-γ (200 U/ml) was added both before and after infection. In cultures treated after infection, IFN-γ and/or supplemental tryptophan were added after infection only, whereas supplemental tryptophan (Trp, 100 μg/ml) was added before infection only.

[b]The values for % R (percentage of cells infected), RI (rickettsiae per infected cell), and NR (rickettsiae per cell) on day 2 are expressed as percentages of the respective values on day 0. Each value represents the mean ± SEM. On day 0, the average values for % R and RI in the various cultures ranged from 78 to 95% and from 4.4 to 9.1, respectively.

[c]The values of NR in the treated cultures on day 2 (expressed as percentages of the values of NR on day 0) are expressed as percentages of the respective control values. Each value represents the mean ± SEM. Values that were significantly different from 100 (as determined by the one-sample *t* test) are indicated as follows: *$p < 0.1$, **$p < 0.05$, ***$p < 0.01$, or ****$p < 0.001$. A significant difference between treated cultures infected with a given rickettsial strain and the corresponding cultures infected with the Madrid E strain (as determined by the two-sample *t* test) is indicated as follows: †$p < 0.1$, ††$p < 0.05$, or †††$p < 0.005$.

TABLE IV

Resistance of *R. prowazekii* Strains 83-2P and 103-2P to IFN-γ in X-Irradiated Human Foreskin Fibroblasts

Rickettsial strain	Cell treatment[a]	No. of expts.	Infection on day 2 (% of value on day 0)[b]			Rickettsial growth (% control)[c]
			% R	RI	NR	
Madrid E	Control	3	106 ± 9	1370 ± 80	1462 ± 166	100
	IFN-γ	3	64 ± 8	119 ± 7	74 ± 5	5 ± 0***
	Trp	3	108 ± 11	1335 ± 115	1415 ± 118	98 ± 6
	Trp + IFN-γ	3	69 ± 6	146 ± 21	100 ± 13	7 ± 0***
83-2P	Control	2	117 ± 1	1411 ± 250	1652 ± 303	100
	IFN-γ	2	99 ± 2	455 ± 105	446 ± 94	27 ± 1**†††
	Trp	2	121 ± 1	1376 ± 204	1670 ± 268	102 ± 3
	Trp + IFN-γ	2	104 ± 4	476 ± 127	505 ± 151	29 ± 4*†
103-2P	Control	2	108 ± 1	1178 ± 107	1275 ± 110	100
	IFN-γ	2	83 ± 2	305 ± 68	250 ± 49	19 ± 2*††
	Trp	2	99 ± 3	1218 ± 143	1199 ± 105	94 ± 0
	Trp + IFN-γ	2	96 ± 4	427 ± 11	410 ± 4	33 ± 3*††

[a] Human foreskin fibroblasts were infected with rickettsiae on day 0. IFN-γ (200 U/ml) and additional tryptophan (Trp, 100 μg/ml) were added to the cultures after infection as indicated.
[b] The values for % R (percentage of cells infected), RI (rickettsiae per infected cell), and NR (rickettsiae per cell) on day 2 are expressed as percentages of the respective values on day 0. Each value represents the mean ± SEM. On day 0, the values of % R and RI (respectively) were 61 ± 3% and 4.0 ± 0.5 in the Madrid E-infected cultures, 56 ± 6% and 4.6 ± 1.3 in the strain 83-2P-infected cultures, and 77 ± 8% and 5.7 ± 1.1 in the strain 103-2P-infected cultures.
[c] The values of NR in the treated cultures on day 2 (expressed as percentages of the values of NR on day 0) are expressed as percentages of the respective control values. Each value represents the mean ± SEM. Values that were significantly different from 100 (as determined by a one-sample *t* test) are indicated as follows: *$p < 0.05$, **$p < 0.01$, or ***$p < 0.001$. A significant difference between treated cultures infected with a given rickettsial strain and the corresponding cultures infected with the Madrid E strain (as determined by the two-sample *t* test) is indicated as follows: †$p < 0.025$, ††$p < 0.01$, or †††$p < 0.001$.

The nitric oxide synthase-independent mechanisms involved in IFN-γ-induced suppression of the growth of *R. prowazekii* in L929 cells have not been defined. However, Gao *et al.*[37] demonstrated that rickettsial DNA and rRNA syntheses (but not rickettsial protein synthesis) are suppressed 12 hr after IFN-γ is added to *R. prowazekii* Madrid E-infected L929 cell cultures. In contrast, rickettsial DNA, rRNA, and protein syntheses are all inhibited 20 hr after IFN-γ is added to *R. prowazekii* Madrid E-infected L929 cell cultures.[37]

2.3. Cytotoxicity of Interferon Treatment and Rickettsial Infection for Host Cells

The effects of IFN-γ on rickettsial infections are not limited to antirickettsial growth inhibitory effects; in addition to such effects, IFN-γ treatment and rickettsial infection in combination have a cytotoxic effect on the host cells. Whereas untreated, nonmultiplying host cells infected with rickettsiae begin to die only after considerable rickettsial growth has occurred, cytotoxic effects on infected host cells treated with IFN-γ are observed at earlier times after infection in host cells with lower rickettsial burdens. It has been suggested that these cytotoxic effects mediated by IFN-γ plus rickettsial infection contribute to damaging tissues in rickettsia-infected hosts.[8,38] Cytotoxicity of IFN-γ treatment and rickettsial infection for the host cells has been observed in studies with: (1) human fibroblasts and *R. prowazekii*,[8,20] (2) mouse L929 cells and *R. prowazekii*,[8,36] (3) mouse macrophagelike RAW264.7 cells and *R. prowazekii*,[39] and (4) BALB/3T3 clone A31 fibroblasts and *R. tsutsugamushi* Gilliam.[40] Although IFN-γ does not inhibit the multiplication of *R. conorii* Casablanca in HEp-2 cells, this cytokine and the rickettsial infection in combination are somewhat cytotoxic for the cells.[31] The combination of IFN-α/β treatment and *R. prowazekii* infection is also cytotoxic for mouse L929 cells.[10] Infection with IFN-α/β-resistant and IFN-γ-resistant isolate *R. prowazekii* 83-2P kills significantly fewer IFN-γ-treated L929 cells than does infection with the IFN-sensitive, parental Madrid E strain.[36] Similarly, the Karp strain of *R. tsutsugamushi* (which is not susceptible to rat IFN-γ in BALB/3T3 clone A31 cells) is not cytotoxic for these IFN-γ-treated host cells, whereas the rat IFN-γ-sensitive Gilliam strain is cytotoxic for rat IFN-γ-treated BALB/3T3 clone A31 cells.[24,40] The mechanisms by which any of these host cells are killed are unknown.

The killing of L929 cells (cytotoxicity) brought about by the combination of *R. prowazekii* infection and IFN-γ treatment is independent of nitric oxide synthase because inhibition of this enzyme does not prevent it.[36] Alterations in the lipid metabolism of *R. prowazekii* Madrid E-infected L929 cell cultures occur when IFN-γ is added to the cultures immediately after infection.[41] Infected, IFN-γ-treated L929 cell cultures exhibit increased phospholipid hydrolysis (and resultant production of free fatty acids) compared with infected, untreated L929 cell cultures. In addition, increased amounts of fatty acids are incorporated into triglycerides in the infected, IFN-γ-treated cultures in comparison with the infected, untreated cultures.[41] Preliminary studies with IFN-resistant *R. prowazekii* strain 60P-infected L929 cells indicated that phospholipid hydrolysis and incorporation of fatty acids into triglycerides did not differ significantly between the untreated and IFN-γ-treated cultures (H. H. Winkler, L. C. Day, R. M. Daugherty, and J. Turco, unpublished data). It is possible that the observed alterations in lipid metabolism contribute to the increased cytotoxicity observed in Madrid E-infected, IFN-γ-treated L929 cells. The absence of these alterations in lipid metabolism in IFN-γ-treated L929 cells infected with IFN-resistant *R. prowazekii* organisms may in turn be correlated with a reduction in the capacity of these rickettsiae to kill the IFN-γ-treated L929 cells.

Macrophagelike RAW264.7 cells pretreated with IFN-γ (or high concentrations of IFN-α/β[23]) are highly susceptible and are killed within hours after infection with viable *R. prowazekii* (Breinl, Madrid E, and IFN-resistant strains derived from Madrid E).[23,39] In contrast, dead *R. prowazekii* organisms are not cytotoxic for the IFN-γ-treated RAW264.7 cells.[39] The cytotoxic capacity of the Breinl strain for the IFN-treated RAW264.7 cells is somewhat higher than that of the Madrid E strain and the IFN-resistant strains.[23] It has been determined that: (1) the presence of exogenous lipopolysaccharide or tumor necrosis factor alpha (TNF-α) is apparently not required for IFN-γ to prepare the RAW264.7 cells for being rapidly killed by *R. prowazekii* (the cytotoxic response)[42]; (2) the *R. prowazekii*-mediated killing of IFN-γ-treated mouse macrophagelike cells does not require the macrophage respiratory burst[38] or the nitric oxide synthase pathway[42]; (3) the *R. prowazekii*-mediated killing of IFN-γ-treated RAW264.7 cells is unlikely to be related to the release of TNF-α from the RAW264.7 cells[42]; (4) growth of the rickettsiae is not required for killing of the IFN-γ-treated RAW264.7 cells to occur, as the killing is not prevented by antirickettsial antibiotics[38]; (5) treatment of the IFN-γ-treated RAW264.7 cells with cytochalasin B (an inhibitor of the induced phagocytosis of rickettsiae[43]) inhibits the killing (J. Turco and H. H. Winkler, unpublished data); and (6) treatment of the rickettsiae with antirickettsial IgG also inhibits the killing.[38] These findings suggest that proper entry and the initial metabolism of the rickettsiae are necessary for the cytotoxicity seen in IFN-γ-treated RAW264.7 cells, but that growth of the rickettsiae is not required.

2.4. Isolation of IFN-Resistant Rickettsiae from *R. prowazekii* Madrid E-Infected L929 Cells

R. prowazekii organisms resistant to IFN-γ have been isolated from strain Madrid E-infected L929 cells after growing the infected cells in the presence of IFN-γ.[44] Two resistant isolates were plaque-purified from independent experiments (strains 427-19 and 87-17). Interestingly, strain 427-19 is also resistant to IFN-α/β treatment in L929 cells.[22,23,44] Growth of *R. prowazekii* Madrid E in untreated L929 cells (which produce IFN-α/β in response to the rickettsial infection) permitted the isolation of additional *R. prowazekii* organisms resistant to both IFN-α/β and IFN-γ.[22] Four resistant isolates were plaque-purified from independent experiments (strains 83-2P, 60P, 103-2P, and 110-1P); and the latter two strains were obtained from L929 cell cultures that had been infected with plaque-purified *R. prowazekii* Madrid E. The mechanisms involved in the resistance of these strains to IFN have not been determined. However, the observation of resistance to IFN-γ in all four strains selected for resistance to IFN-α/β and the observation of resistance to IFN-α/β in only one of the two strains selected for resistance to IFN-γ suggest that there is not only a common rickettsial target or activity but also some rickettsial target or activity that contributes to suppression of rickettsial growth in IFN-γ-treated, but not in IFN-α/β-treated, L929 cells.

A very notable biological correlate of virulence in *R. prowazekii* is the ability to grow well in macrophages. Specifically, the virulent Breinl strain grows well in untreated cultures of human monocyte-derived macrophages[45] and untreated cultures of mouse macrophagelike RAW264.7 cells,[46] whereas the avirulent Madrid E strain does not. The question of a possible correlation between resistance to IFNs and virulence in *R. prowazekii* was raised when we discovered that all of the IFN-resistant strains derived from the avirulent Madrid E strain are similar to the virulent Breinl strain (and unlike the avirulent Madrid E strain) in their ability to grow very well in untreated macrophagelike RAW264.7 cells.[23]

Experiments revealed that the virulent Breinl strain of *R. prowazekii* is sensitive to IFN-γ but resistant to IFN-α/β in L929 cells (compared with the avirulent Madrid E strain). Therefore, resistance to IFN-γ does not appear to correlate with virulence in *R. prowazekii*,[23] but it may be that resistance to IFN-α/β (in L929 cells) is correlated with virulence. The possibility of a correlation between resistance to IFNs in mouse L929 cells and virulence was further explored by evaluating the IFN sensitivity of *R. prowazekii* EVir, a virulent strain derived by Balayeva and Nikolskaya,[47] who passed the avirulent Madrid E strain in the lungs of white mice. Strain EVir is resistant to both IFN-α/β and IFN-γ in L929 cells; in addition, strain EVir (like the virulent Breinl strain and all of the IFN-resistant *R. prowazekii* strains) grows very well in untreated RAW264.7 cells.[23] Like strain 60P (a Madrid E-derived strain selected for resistance to IFN-α/β), strain EVir is also a very poor inducer of IFN-α/β in L929 cells.[23,48] As Rodionov *et al.*[49] had determined that strains EVir and Breinl are similar to each other and different from strain Madrid E in the lysine methylation profiles exhibited by their surface protein antigens, we examined and compared lysine methylation in *R. prowazekii* 83-2P (an IFN-α/β- and IFN-γ-resistant strain selected for resistance to IFN-α/β), *R. prowazekii* 87-17 (an IFN-γ-resistant, IFN-α/β-sensitive strain), *R. prowazekii* EVir, and *R. prowazekii* Madrid E. The results indicated similarity in lysine methylation among strains 83-2P, 87-17, and EVir; however, lysine methylation in these three strains differed significantly from that observed in strain Madrid E.[48] Thus, the virulent *R. prowazekii* strain EVir and the IFN-resistant *R. prowazekii* strains (which have not been evaluated for virulence) are similar in several properties.

The cytokine-resistant *R. prowazekii* isolates have retained their cytokine resistance after being plaque-purified two times in untreated cultured cells, and then being grown in the yolk sacs of embryonated chicken eggs.[22,23,44] These observations suggest that the cytokine resistance of these isolates is associated with a genotypic change in (rather than a phenotypic adaptation of) the rickettsiae. On the other hand, the IFN-γ-resistant rickettsiae derived from *R. tsutsugamushi* Gilliam by Hanson[40] lost their resistance to IFN-γ after a few passages in untreated, cultured cells. These findings suggest that the IFN-γ resistance of these *R. tsutsugamushi* organisms represents an adaptive response of the bacteria. In both cases the isolation of cytokine-resistant rickettsiae is of considerable interest, as cytokine resistance may be related to the ability of rickettsiae to avoid destruction and persist in their hosts after the latter have recovered.

3. STUDIES INVOLVING TUMOR NECROSIS FACTORS

3.1. Induction of Tumor Necrosis Factors by Rickettsiae

Production of tumor necrosis factor alpha (TNF-α) is induced in cultures of human monocyte-derived macrophages infected with *R. conorii* or Israeli spotted fever rickettsiae,[50] in cultures of mouse peritoneal macrophages and mouse macrophagelike P388D1 cells infected with *R. conorii* Malish 7,[51] and in cultures of mouse macrophagelike RAW264.7 cells infected with *R. prowazekii* (Madrid E and Breinl strains).[42] Rickettsial viability is not required for induction of TNF-α production, as both heat-killed *R. conorii*[51] and heat-killed *R. prowazekii* organisms[42] induce the production of TNF-α. TNF-α is also detected in culture supernatant fluids collected after incubation of human monocyte-derived macrophages with *R. conorii*-infected HEp-2 cell cultures.[52] *R. tsutsugamushi* Karp organisms (whether viable or heat-killed) do not induce the production of detectable TNF-α in cultures of mouse peritoneal macrophages and macrophagelike P388D1 cells.[51] However, immune mouse spleen cells incubated with heat-killed *R. tsutsugamushi* and immune mouse spleen cells incubated with heat-killed *R. conorii* both produce TNF.[51] TNF is produced during infection of mice with *R. australis*,[53] and it is detected in the sera of immune mice given the appropriate rickettsial antigen (*R. conorii* or *R. tsutsugamushi*).[51] TNF-α is also produced in mice[15] and humans[54] infected with *R. conorii*. Kern *et al.*[54] have demonstrated that the levels of the soluble TNF receptors (sTNFR-p55 and sTNFR-p75) are increased in the plasma of humans with acute Mediterranean spotted fever. They reported that the levels of both soluble TNF receptors are higher in patients with severe rather than mild illness.

3.2. Effects of Tumor Necrosis Factors on Rickettsial Growth

TNF-α restricts the multiplication of *R. conorii* Casablanca in human HEp-2 cells, and this inhibitory effect of TNF-α does not appear to require host cell protein synthesis and is partially alleviated by supplementation of the culture medium with additional tryptophan.[31] TNF-α also suppresses the growth of *R. conorii* Malish 7 in mouse L929 cells,[30] and Feng *et al.*[15] demonstrated the importance of TNF-α in protecting challenged mice from lethality caused by *R. conorii* Malish 7 infections. TNF-α suppresses the growth of *R. prowazekii* Madrid E and Breinl[36] in L929 cells; and the growth of *R. tsutsugamushi* Karp in mouse macrophages and the mouse cell line C3H/10T1/2 clone 8.[32] The suppression of growth of *R. conorii* by TNF-α in L929 cells is alleviated by inhibition of nitric oxide synthase.[30] However, TNF-α-induced inhibition of *R. prowazekii* Madrid E in L929 cells is not alleviated and TNF-α-induced inhibition of *R. prowazekii* Breinl is slightly alleviated by inhibitors of nitric oxide synthase.[36] Therefore, in L929 cells infected with the Madrid E or Breinl strains, a nitric oxide synthase-independent mechanism plays a significant role in mediating the antirickettsial effects of TNF-α.[36] It is possible that *R. conorii* Malish 7 is resistant to the TNF-α-induced, nitric oxide synthase-independent mechanism that restricts the growth of *R. prowazekii* in L929 cells. However, as discussed in Section 2.2, the possibility that *R. conorii* Malish 7 is sensitive to both nitric oxide synthase-independent as well as nitric oxide synthase-dependent, cytokine-induced, inhibitory mechanisms cannot be eliminated. Our finding that IFN-α/β-resistant, IFN-γ-resistant *R. prowazekii* strains 83-2P and EVir are also resistant to TNF-α (compared with the Madrid E strain) suggests that there is some commonality among the nitric oxide synthase-independent, antirickettsial effects induced by IFN-α/β, IFN-γ, and TNF-α.[36,48]

3.3. Cytotoxicity of Tumor Necrosis Factor Treatment and Rickettsial Infection for Host Cells

We noted that TNF-α and *R. prowazekii* in combination (like IFN-γ treatment and *R. prowazekii* in combination) have a cytotoxic effect on mouse L929 cells,[36] and that this toxicity is not alleviated by the inhibition of nitric oxide synthase.[36] Significantly fewer TNF-α-treated L929 cells are killed by the TNF-α- and IFN-resistant isolate *R. prowazekii* 83-2P compared with the TNF-α- and IFN-sensitive, parental Madrid E strain.[36] In contrast, Manor and Sarov[31] did not observe such toxicity in their experiments with TNF-α-treated, *R. conorii*-infected HEp-2 cells; Feng and Walker[30] did not report such toxicity in their studies with *R. conorii*-infected, TNF-α-treated L929 cells, and Geng and Jerrells[32] did not note toxicity in their studies with TNF-α-treated, *R. conorii*-infected mouse macrophages and C3H/10T1/2 clone 8 cells. Pretreatment of macrophagelike RAW264.7 cells with TNF-α alone does not prepare these cells to be rapidly killed after they are infected with *R. prowazekii*.[42]

4. STUDIES INVOLVING IFN-γ PLUS TNF-α

In studies with human HEp-2 cells and *R. conorii* Casablanca, Manor and Sarov[31] found that IFN-γ (which does not by itself inhibit growth of *R. conorii* in this system) enhances the inhibitory effect of TNF-α on rickettsial growth. They also observed that treatment of HEp-2 cells with both cytokines (IFN-γ plus TNF-α) along with *R. conorii* infection has a pronounced cytotoxic effect on the cells (a cytotoxic effect much greater than the toxicity observed in infected IFN-γ-treated HEp-2 cells). Interestingly, both effects observed with IFN-γ plus TNF-α are partially alleviated by supplemental tryptophan.[31] Geng and Jerrells[32] stated that IFN-γ does not consistently increase the inhibitory effect of TNF-α on growth of the Karp strain of *R. tsutsugamushi* in mouse C3H/10T1/2 cells. In studies with *R. conorii* Malish 7 and mouse L929 cells, Feng and Walker[30] observed that production of nitric oxide correlates with inhibition of growth of *R. conorii* in L929 cells treated with IFN-γ alone, TNF-α alone, or IFN-γ plus TNF-α. Synergistic induction of nitric oxide synthesis by IFN-γ plus TNF-α is associated with synergistic inhibition of growth of *R. conorii*.[30]

Other experiments conducted by Feng *et al.*[15] have clearly demonstrated the importance of both IFN-γ and TNF-α in allowing mice to recover from infection with *R. conorii* Malish 7. Whereas the infected, sham-depleted mice in their study all recovered, infected mice depleted of IFN-γ, TNF-α, or both cytokines (through the administration of neutralizing antibodies) all died. Feng *et al.*[15] observed markedly increased numbers of viable rickettsiae per gram of liver tissue in the infected mice that had been depleted of either IFN-γ or TNF-α (compared with the numbers observed in infected, sham-depleted animals), and in mice depleted of both cytokines, the increase was even more dramatic.[15] Measurements indicated that urinary nitrate (an indicator of nitric oxide synthesis) was consistently increased in sham-depleted, *R. conorii*-infected mice, but not in mice depleted of IFN-γ, TNF-α, or both cytokines[15]; these data suggest that nitric oxide synthesis correlates with protection of the mice. Although these experiments demonstrate the importance of both IFN-γ and TNF-α as *in vivo* host defenses against *R. conorii* infections in mice and suggest that the nitric oxide synthase pathway plays an important role in protecting the mice, it is also possible that nitric oxide synthase-independent mechanisms are involved.

A nitric oxide synthase-independent mechanism plays a significant role in cytokine-induced inhibition of growth of *R. prowazekii* Madrid E and Breinl strains in L929 cells: in infected L929 cells treated with IFN-γ alone, TNF-α alone, or IFN-γ plus TNF-α, inhibitors of nitric oxide synthase have little or no mitigating effect on the observed inhibition of rickettsial growth.[36] However, in L929 cell cultures infected with IFN-α/β-, IFN-γ-, and TNF-α-resistant *R. prowazekii* strain 83-2P, IFN-γ and TNF-α in combination have a synergistic inhibitory effect on rickettsial growth that is relieved by inhibition of nitric oxide synthesis.[36] It follows that strain 83-2P is resistant to the cytokine-induced, nitric oxide synthase-independent inhibitory mechanism, but sensitive to the nitric oxide synthase-dependent inhibitory mechanism induced by IFN-γ plus TNF-α. *R. prowazekii* strain EVir is similar to strain 83-2P in being resistant to cytokine-induced nitric oxide synthase-independent inhibition but sensitive to cytokine-induced nitric oxide synthase-dependent inhibition in L929 cells.[48]

The cytotoxicity observed in *R. prowazekii* Madrid E-infected L929 cell cultures treated with IFN-γ plus TNF-α was not markedly greater than the cytotoxicity observed in infected L929 cell cultures treated with TNF-α alone.[36] However infected cultures treated with IFN-γ plus TNF-α and infected cultures treated with TNF-α alone contained somewhat higher percentages of dead L929 cells than infected cultures treated with IFN-γ alone. In no case were the cytotoxic effects on the L929 cells alleviated by inhibition of nitric oxide synthase.[36] The rapid cytotoxic response has not been evaluated in macrophagelike RAW264.7 cells pretreated with IFN-γ plus TNF-α and then infected with *R. prowazekii*.

5. ANTIRICKETTSIAL EFFECTS OF CRUDE LYMPHOKINES AND THE ROLE OF IFN-γ AND OTHER CYTOKINES IN MEDIATING THESE EFFECTS

5.1. Studies with Crude Lymphokines

In vitro and *in vivo* studies involving several *Rickettsia* species and crude lymphokine preparations were reviewed previously.[7-9] Crude lymphokine preparations restrict the growth of *R. tsutsugamushi* Gilliam and Karp in mouse macrophages[32,55-58]; the growth of *R. conorii* Casablanca[28] in mouse macrophages; the growth of *R. prowazekii* Breinl in mouse macrophagelike RAW264.7 cells,[39] human monocyte-derived macrophages,[20] and human endothelial cells[20]; the multiplication of *R. prowazekii* Madrid E and Breinl in mouse L929 cells[59] and human fibroblasts[20,59]; and the multiplication of *R. conorii* in mouse L929 cells.[59] Lymphokine treatment of macrophages obtained from *R. akari*-resistant mouse strains (C3H/HeN, C57BL/10J, and BALB/cN) restricts the growth of *R. akari* within the macrophages.[60] In contrast, in macrophages obtained from *R. akari*-susceptible mouse strains (C3H/HeJ, C57BL/10SnCR, and A/J), lymphokine-induced restriction of the growth of *R. akari* is markedly diminished.[60] In studies with *R. tsutsugamushi* Gilliam, however, lymphokines similarly restrict rickettsial growth in macrophages from resistant BALB/c mice and susceptible C3H/He mice.[56]

In addition to the antirickettsial effects described, cytotoxic effects on human fibroblasts[20] and mouse macrophagelike RAW264.7 cells[39] brought about by the combination of lymphokine treatment and *R. prowazekii* infection are observed. Crude lymphokines and *R. tsutsugamushi* Gilliam infection are also toxic for mouse macrophages.[61] Finally, lymphokine treatment of mouse macrophages is associated with suppression of the ability of *R. tsutsugamushi* Gilliam and *R. akari* Kaplan to initially infect the macrophages[58,60]; and treatment of mouse macrophagelike RAW264.7 cells and mouse L929 cells with high concentrations of crude lymphokines is followed by inhibition of the ability of *R. prowazekii* Madrid E and Breinl to initially infect the cells.[8,39]

5.2. Role of IFN-γ and Other Factors in Crude Lymphokines

The results of studies conducted with *R. prowazekii*[27,28,39] and *R. conorii*[28] are consistent with the idea that IFN-γ in the crude lymphokines plays a major role in limiting rickettsial growth in lymphokine-treated, cultured macrophages and fibroblasts. For example, treatment of crude lymphokines with IFN-γ-neutralizing monoclonal antibody conjugated to Sepharose beads eliminated the ability of the lymphokines to restrict the growth of *R. prowazekii* in mouse L929 cells and the growth of *R. conorii* in mouse macrophages.[28] In addition, as discussed in Section 2, recombinant IFN-γ itself restricts the multiplication of *R. prowazekii* in L929 cells[27] and the multiplication of *R. conorii* in macrophages.[28] On the other hand, treatment of *R. prowazekii*-infected L929 cells with various concentrations of IFN-γ as either crude lymphokines or recombinant IFN-γ showed that, at high concentrations of IFN-γ, the observed, maximal inhibition of rickettsial growth was slightly greater when the L929 cells were treated with crude lymphokines rather than recombinant IFN-γ.[27]

Experiments conducted by Nacy *et al.*[55] with mouse macrophages and *R. tsutsugamushi* first provided evidence that there is more than one antirickettsial factor present in crude lymphokine preparations. Chromatographic analysis of crude lymphokine preparations on a column of Sephadex G-200, followed by assay of the resultant fractions for ability to induce macrophage killing of intracellular *R. tsutsugamushi* organisms, yielded three separate peaks of antirickettsial activity.[55] However, only one of the peaks (the peak later shown to have IFN-γ activity) had the ability to cause suppression of the initial *R. tsutsugamushi* infection in treated macrophages.[55]

Studies of lymphokine-induced suppression of the initial *R. prowazekii* infection in L929 cells suggest a possible requirement for a factor or factors in addition to IFN-γ. Although high concentrations of IFN-γ are like crude lymphokines in inhibiting the ability of macrophagelike RAW264.7 cells to be initially infected with *R. prowazekii*,[39] high concentrations of IFN-γ differ from crude lymphokines because the former do not inhibit the ability of L929 cells to be initially infected.[8,62] However, neutralization of the IFN-γ in the crude lymphokines prevents the lymphokines from causing inhibition of the initial rickettsial infection in L929 cells.[8,62] A plausible explanation for these data is that lymphokine-induced inhibition of the initial *R. prowazekii* infection requires both IFN-γ and some other factor(s).

5.3. Role of Nitric Oxide Production

Experimental evidence has indicated that: (1) lymphokine-induced suppression of the initial *R. prowazekii* infection in L929 cells is dependent on nitric oxide synthase; (2) IFN-γ alone (even at high concentrations) does not cause suppression of the initial rickettsial infection in L929 cells because it does not induce the production of nitric oxide by these cells; and (3) TNF-α and IFN-γ in combination effectively induce the production of nitric oxide in L929 cells and inhibit the initial *R. prowazekii* infection in the cells.[63] Because treatment of isolated *R. prowazekii* organisms with nitric oxide gas inhibits their ability to infect L929 cells,[64] and because addition of oxyhemoglobin (a scavenger of extracellular nitric oxide) to cytokine-treated cultures prevents the suppression of the initial *R. prowazekii* infection, nitric oxide released from appropriately treated L929 cells is likely to be responsible for inhibiting the ability of *R. prowazekii* organisms to initially infect the L929 cells.[63]

Treatment of isolated *R. prowazekii* organisms with nitric oxide gas also inhibits their ability to infect and kill IFN-γ-treated RAW264.7 cells.[64] Suppression of the initial *R. prowazekii* infection and suppression of *R. prowazekii*-mediated rapid killing of IFN-γ-treated RAW264.7 cells are observed in RAW264.7 cell cultures pretreated with any of the following: (1) high concentrations of crude lymphokines,[39] (2) high concentrations of IFN-γ,[39] or (3) bacterial lipopolysaccharide plus low concentrations of IFN-γ.[42,63] The importance of nitric oxide released from the RAW264.7 cells in these events is suggested by our finding that addition of oxyhemoglobin prevents both the suppression of the initial rickettsial infection and the suppression of the *R. prowazekii*-mediated rapid killing of the RAW264.7 cells in cultures treated with lipopolysaccharide plus IFN-γ.[63] It would be interesting to determine if high concentrations of IFN-γ induce the synthesis by RAW264.7 cells of a factor(s) that acts together with IFN-γ to induce the production of large amounts of nitric oxide by the RAW264.7 cells. Furthermore, in light of the effectiveness of high concentrations of crude lymphokines (but not high concentrations of IFN-γ) at inducing the production of nitric oxide by L929 cells, it would be worthwhile to conduct experiments with nitric oxide synthase inhibitors to test the hypothesis that the slightly greater levels of inhibition of growth of *R. prowazekii* observed in L929 cell cultures treated with high concentrations of crude lymphokines[27] are dependent on nitric oxide synthase.

6. EVIDENCE FOR KILLING OF INTRACELLULAR RICKETTSIAE IN CYTOKINE-TREATED HOST CELLS

6.1. Studies with Cultured Cells

The role of cytokines as effectors in host defense against rickettsial infections is influenced by whether the cytokines are rickettsicidal or rickettsistatic. At least three types of observations support the idea that many intracellular rickettsiae are apparently killed in cytokine-treated, cultured cells. First, the number of rickettsiae per infected cell in cytokine-treated cell cultures is sometimes observed to fall (this is seen in nondividing host cells and is not an artifact of host cell division). Second, reduction in the percentage of nondividing host cells infected with rickettsiae suggests that the rickettsial infection has been cleared from these cells. Third, decreases in the number of viable rickettsiae per culture have been detected by plaque assays. Specifically, Nacy and Meltzer[58] provided convincing evidence that many intracellular *R. tsutsugamushi* organisms are killed in lymphokine-treated mouse macrophages. In experiments in which the macrophages were first infected and then treated with crude lymphokines, both the percentages of host cells infected with rickettsiae and the numbers of viable rickettsiae (detected by plaque assays) in the lymphokine-treated cultures decreased dramatically (relative to those in untreated macrophages) within the first 24 hr.[58] Decreases over time in the percentages of nondividing host cells infected have also been observed in lymphokine-treated cultures of: human monocyte-derived macrophages infected with *R. prowazekii* Breinl[20]; human fibroblasts infected with *R. prowazekii* Breinl[20]; and mouse L929 cells infected with *R. prowazekii* Madrid E or *R. prowazekii* Breinl.[59] In addition, decreases over time in the number of *R. prowazekii* organisms per infected cell are observed in nonmultiplying L929 cell cultures treated with lymphokines both before and after infection.[59] In IFN-γ-treated cultures, decreases (over time) in the percentages of nondividing host cells infected have been noted in experiments with: (1) mouse L929 cells[27] and human MRC-5 fibroblasts (Table II) infected with *R. prowazekii* and (2) mouse BALB/3T3 clone A31 cells infected with *R. tsutsugamushi* Gilliam.[40] These findings provide evidence that rickettsiae are cleared from some of the cytokine-treated, infected host cells. [It is important to note, however, that in some (but not all) instances, interpretation of the data is complicated by the observation of cytotoxic effects on the cytokine-treated, infected host cells.]

6.2. Studies of *R. conorii*-Infected Mice

The results of studies of cytokine-depleted and nondepleted mice infected with *R. conorii* Malish 7[15] (discussed in part in Section 4) are consistent with the hypothesis that IFN-γ and TNF-α are involved in mediating the killing of intracellular rickettsiae. Immunohistologic and ultrastructural studies showed that, several days after infection of the nondepleted animals, rickettsiae were difficult to find and most of the rickettsiae observed within the macrophages of the livers appeared to be located in phagolysosomes.[15] In contrast, in the cytokine-depleted animals, there were more rickettsiae and a greater proportion of the rickettsiae appeared morphologically intact.[15]

7. CONCLUDING REMARKS

Both *in vivo* and *in vitro* studies have made significant contributions to current knowledge about the influence of interferons and tumor necrosis factors on rickettsial infections. Studies of rickettsial infections in animal models have established the importance of IFN-γ and TNF-α as antirickettsial host defenses; and studies of rickettsial infections in cultured cells have provided information about: (1) how these cytokines modify the interactions between rickettsiae and their host cells and (2) the involvement of both nitric oxide synthase-dependent and -independent mechanisms in the observed alterations. Inhibition of rickettsial growth, killing of intracellular rickettsiae, and cytotoxic effects on the host cells have been observed when host cells are both infected with rickettsiae and treated with cytokines. There is evidence for the involvement of both nitric oxide synthase-dependent and -independent mechanisms in the cytokine-induced effects on rickettsial growth and viability. However, experimental data argue against the involvement of a nitric oxide synthase-dependent mechanism in the cytotoxic effects on host cells that are brought about by the combination of rickettsial infection and cytokine treatment. The occurrence of cytotoxicity in rickettsia-infected, cytokine-treated host cells has led to the suggestion that cytokines contribute in some way to the pathogenesis of rickettsial infections as well as to antirickettsial host defense. On the other hand, inhibition of the initial rickettsial infection in cytokine-treated host cells is entirely dependent on nitric oxide synthase, and exposure of isolated rickettsiae to nitric oxide gas kills them.

ACKNOWLEDGMENTS. We are grateful to the following colleagues for generously providing cell lines used in the experiments reported in Table II: Dr. Esther Manor, HEp-2 cells; Dr. Charles E. Samuel, GM2767A cells; and Dr. Ganes C. Sen, RD114 and RD114-A2 cells. Recombinant human IFN-γ derived from *Escherichia coli* was a gift from Genentech, Inc., South San Francisco, California; and recombinant human IFN-β_{ser} was a gift from Triton Biosciences Inc., Alameda, California.

REFERENCES

1. Weiss, E., and Moulder, J. W., 1984, Genus I. Rickettsia, in: *Bergey's Manual of Systematic Bacteriology*, Volume 1 (N. R. Krieg and J. G. Holt, eds.), Williams & Wilkins, Baltimore, pp. 688–698.
2. Tamura, A., Ohashi, N., Urakami, H., and Miyamura, S., 1995, Classification of *Rickettsia tsutsugamushi* in a new genus, *Orientia* gen. nov., as *Orientia tsutsugamushi* comb. nov., *Int. J. Syst. Bacteriol.* **45**:589–591.
3. Higgins, J. A., Radulovic, S., Schriefer, M. E., and Azad, A. F., 1996, *Rickettsia felis*—A new species of pathogenic rickettsia isolated from cat fleas, *J. Clin. Microbiol.* **34**:671–674.
4. Stothard, D. R., and Fuerst, P. A., 1995, Evolutionary analysis of the spotted fever and typhus groups of *Rickettsia* using 16S rRNA gene sequences, *Syst. Appl. Microbiol.* **18**:52–61.
5. Nacy, C. A., Meltzer, M. S., Jerrells, T. R., and Byrne, G. I., 1989, Chlamydial and rickettsial infections, in: *Natural Immunity* (D. S. Nelson, ed.), Academic Press Australia, Marrickville, NSW pp. 587–612.
6. Winkler, H. H., and Turco, J., 1984, Role of lymphokine (gamma interferon) in host defense against *Rickettsia prowazekii*, in: *Microbiology—1984* (L. Leive and D. Schlessinger, eds.), American Society for Microbiology, Washington, DC, pp. 273–276.

7. Winkler, H. H., and Turco, J., 1988, *Rickettsia prowazekii* and the host cell: Entry, growth and control of the parasite, *Curr. Top. Microbiol. Immunol.* **138:**81–107.

8. Turco, J., and Winkler, H. H., 1988, Interactions between *Rickettsia prowazekii* and cultured host cells: Alterations induced by gamma interferon, in: *Interferon and Nonviral Pathogens* (G. I. Byrne and J. Turco, eds.), Dekker, New York, pp. 95–115.

9. Winkler, H. H., and Turco, J., 1993, Rickettsiae and macrophages, in: *Macrophage–Pathogen Interactions* (B. S. Zwilling and T. K. Eisenstein, eds.), Dekker, New York, pp. 401–414.

10. Turco, J., and Winkler, H. H., 1990, Interferon-α/β and *Rickettsia prowazekii:* Induction and sensitivity, *Ann. N.Y. Acad. Sci.* **590:**168–186.

11. Kohno, S., Kohase, M., Sakata, H., Shimizu, Y., Hikita, M., and Shishido, A., 1970, Production of interferon in primary chick embryonic cells infected with *Rickettsia mooseri, J. Immunol.* **105:**1553–1558.

12. Hopps, H. E., Kohno, S., Kohno, M., and Smadel, J. E., 1964, Production of interferon in tissue cultures infected with *Rickettsia tsutsugamushi, Bacteriol. Proc.* **1964:**115–116.

13. Kazar, J., 1966, Interferon-like inhibitor in mouse sera induced by rickettsiae, *Acta Virol.* **10:**277.

14. Li, H., Jerrells, T. R., Spitalny, G. L., and Walker, D. H., 1987, Gamma interferon as a crucial host defense against *Rickettsia conorii* in vivo, *Infect. Immun.* **55:**1252–1255.

15. Feng, H.-M., Popov, V. L., and Walker, D. H., 1994, Depletion of gamma interferon and tumor necrosis factor alpha in mice with *Rickettsia conorii*-infected endothelium: Impairment of rickettsicidal nitric oxide production resulting in fatal, overwhelming rickettsial disease, *Infect. Immun.* **62:**1952–1960.

16. Walker, D. H., Popov, V. L., Wen, J., and Feng, H.-M., 1994, *Rickettsia conorii* infection of C3H/HeN mice. A model of endothelial-target rickettsiosis, *Lab. Invest.* **70:**358–368.

17. MacMillan, J. G., Rice, R. M., and Jerrells, T. R., 1985, Development of antigen-specific cell-mediated immune responses after infection of cynomolgus monkeys (*Macaca fascicularis*) with *Rickettsia tsutsugamushi, J. Infect. Dis.* **152:**739–749.

18. Palmer, B. A., Hetrick, F. M., and Jerrells, T. R., 1984, Production of gamma interferon in mice immune to *Rickettsia tsutsugamushi, Infect. Immun.* **43:**59–65.

19. Palmer, B. A., Hetrick, F. M., and Jerrells, T. R., 1984, Gamma interferon production in response to homologous and heterologous strain antigens in mice chronically infected with *Rickettsia tsutsugamushi, Infect. Immun.* **46:**237–244.

20. Wisseman, C. L., Jr., and Waddell, A., 1983, Interferonlike factors from antigen- and mitogen-stimulated human leukocytes with antirickettsial and cytolytic actions on *Rickettsia prowazekii*-infected human endothelial cells, fibroblasts, and macrophages, *J. Exp. Med.* **157:**1780–1793.

21. Kazar, J., Krautwurst, P. A., and Gordon, F. B., 1971, Effect of interferon and interferon inducers on infections with a nonviral intracellular microorganism, *Rickettsia akari, Infect. Immun.* **3:**819–824.

22. Turco, J., and Winkler, H. H., 1990, Selection of alpha/beta interferon- and gamma interferon-resistant rickettsiae by passage of *Rickettsia prowazekii* in L929 cells, *Infect. Immun.* **58:**3279–3285.

23. Turco, J., and Winkler, H. H., 1991, Comparison of properties of virulent, avirulent, and interferon-resistant *Rickettsia prowazekii* strains, *Infect. Immun.* **59:**1647–1655.

24. Hanson, B., 1991, Comparative susceptibility to mouse interferons of *Rickettsia tsutsugamushi* strains with different virulence in mice and of *Rickettsia rickettsii, Infect. Immun.* **59:**4134–4141.

25. Groves, M. G., and Osterman, J. V., 1978, Host defenses in experimental scrub typhus: Genetics of natural resistance to infection, *Infect. Immun.* **19:**583–588.

26. Turco, J., and Winkler, H. H., 1986, Gamma interferon-induced inhibition of the growth of *Rickettsia prowazekii* in fibroblasts cannot be explained by the degradation of tryptophan or other amino acids, *Infect. Immun.* **53:**38–46.

27. Turco, J., and Winkler, H. H., 1983, Cloned mouse interferon-gamma inhibits the growth of *Rickettsia prowazekii* in cultured mouse fibroblasts, *J. Exp. Med.* **158:**2159–2164.

28. Jerrells, T. R., Turco, J., Winkler, H. H., and Spitalny, G. L., 1986, Neutralization of lympho-kine-mediated antirickettsial activity of fibroblasts and macrophages with monoclonal antibody specific for murine interferon gamma, *Infect. Immun.* **51:**355–359.

29. Keysary, A., Oron, C., Rosner, M., and Weissman, B. A., 1995, The involvement of l-arginine-nitric oxide pathway in the anti-rickettsial activity of macrophagelike cells, in: *Biochemical, Pharmacological, and Clinical Aspects of Nitric Oxide* (B. A. Weissman, N. Allon, and S. Shapira, eds.), Plenum Press, New York, pp. 111–115.

30. Feng, H.-M., and Walker, D. H., 1993, Interferon-γ and tumor necrosis factor-α exert their antirickettsial effect via induction of synthesis of nitric oxide, *Am. J. Pathol.* **143:**1016–1023.

31. Manor, E., and Sarov, I., 1990, Inhibition of *Rickettsia conorii* growth by recombinant tumor necrosis factor alpha: Enhancement of inhibition by gamma interferon, *Infect. Immun.* **58:**1886–1890.

32. Geng, P., and Jerrells, T. R., 1994, The role of tumor necrosis factor in host defense against scrub typhus rickettsiae. I. Inhibition of growth of *Rickettsia tsutsugamushi*, Karp strain, in cultured murine enbryonic cells and macrophages by recombinant tumor necrosis factor-alpha, *Microbiol. Immunol.* **38:**703–711.

33. Samuel, C. E., and Knutson, G. S., 1983, Mechanism of interferon action: Human leukocyte and immune interferons regulate the expression of different genes and induce different antiviral states in human amnion U cells, *Virology,* **130:**474–484.

34. Kumar, R., Tiwari, R. K., Kusari, J., and Sen, G. C., 1987, Clonal derivatives of the RD-114 cell line differ in their antiviral and gene-inducing responses to interferons, *J. Virol.* **61:**2727–2732.

35. Pfefferkorn, E. R., 1984, Interferon-γ blocks the growth of *Toxoplasma gondii* in human fibroblasts by inducing the host cells to degrade tryptophan, *Proc. Natl. Acad. Sci. USA* **81:**908–912.

36. Turco, J., and Winkler, H. H., 1993, Role of the nitric oxide synthase pathway in the inhibition of growth of interferon-sensitive and interferon-resistant *Rickettsia prowazekii* strains in L929 cells treated with tumor necrosis factor alpha and gamma interferon, *Infect. Immun.* **61:**4317–4325.

37. Gao, Q., Turco, J., and Winkler, H. H., 1993, Synthesis of DNA, rRNA, and protein by *Rickettsia prowazekii* growing in untreated or gamma interferon-treated mouse L929 cells, *Infect. Immun.* **61:**2383–2389.

38. Turco, J., Keysary, A., and Winkler, H. H., 1989, Interferon-γ and rickettsia-induced killing of macrophage-like cells is inhibited by anti-rickettsial antibodies and does not require the respiratory burst, *J. Interferon Res.* **9:**615–629.

39. Turco, J., and Winkler, H. H., 1984, Effect of mouse lymphokines and cloned mouse interferon-gamma on the interaction of *Rickettsia prowazekii* with mouse macrophage-like RAW264.7 cells, *Infect. Immun.* **45:**303–308.

40. Hanson, B., 1991, Susceptibility of *Rickettsia tsutsugamushi* Gilliam to gamma interferon in cultured mouse cells, *Infect. Immun.* **59:**4125–4133.

41. Winkler, H. H., Day, L. C., Daugherty, R. M., and Turco, J., 1993, Effect of gamma interferon on phospholipid hydrolyis and fatty acid incorporation in L929 cells infected with *Rickettsia prowazekii, Infect. Immun.* **61:**3412–3415.

42. Turco, J., and Winkler, H. H., 1994, Relationship of tumor necrosis factor alpha, the nitric oxide synthase pathway, and lipopolysaccharide to the killing of gamma interferon-treated macrophagelike RAW264.7 cells by *Rickettsia prowazekii, Infect. Immun.* **62:**2568–2574.

43. Walker, T. S., and Winkler, H. H., 1978, Penetration of cultured mouse fibroblasts (L cells) by *Rickettsia prowazekii, Infect. Immun.* **22:**200–208.

44. Turco, J., and Winkler, H. H., 1989, Isolation of *Rickettsia prowazekii* with reduced sensitivity to gamma interferon, *Infect. Immun.* **57:**1765–1772.

45. Gambrill, M. R., and Wisseman, C. L., Jr., 1973, Mechanisms of immunity in typhus infection. II. Multiplication of typhus rickettsiae in human macrophage cell cultures in the nonimmune system: Influence of virulence of rickettsial strains and of chloramphenicol, *Infect. Immun.* **8:**519–527.

46. Turco, J., and Winkler, H. H., 1982, Differentiation between virulent and avirulent strains of *Rickettsia prowazekii* by macrophage-like cell lines, *Infect. Immun.* **35:**783–791.

47. Balayeva, N. M., and Nikolskaya, V. N., 1973, Increased virulence of the E vaccine strain of *Rickettsia prowazekii* passaged in the lungs of white mice and guinea pigs, *J. Hyg. Epidemiol. Microbiol. Immunol.* **17:**11–20.

48. Turco, J., and Winkler, H. H., 1994, Cytokine sensitivity and methylation of lysine in *Rickettsia prowazekii* EVir and interferon-resistant *R. prowazekii* strains, *Infect. Immun.* **62:**3172–3177.

49. Rodionov, A. V., Eremeeva, M. E., and Balayeva, N. M., 1991, Isolation and partial characterization of the M_r 100 kD protein from *Rickettsia prowazekii* strains of different virulence, *Acta Virol.* **35:**557–565.

50. Manor, E., and Sarov, I., 1990, Tumor necrosis factor alpha and prostaglandin E_2 production by human monocyte-derived macrophages infected with spotted fever group rickettsiae, *Ann. N.Y. Acad. Sci.* **590:**157–167.

51. Jerrells, T. R., and Geng, P., 1994, The role of tumor necrosis factor in host defense against scrub typhus rickettsiae. II. Differential induction of tumor necrosis factor-alpha production by *Rickettsia tsutsugamushi* and *Rickettsia conorii*, *Microbiol. Immunol.* **38:**713–719.

52. Manor, E., 1991, The effect of monocyte-derived macrophages on the growth of *Rickettsia conorii* in permissive cells, in: *Rickettsiae and Rickettsial Diseases, Proceedings of the Fourth International Symposium* (J. Kazar and D. Raoult, eds.), Publishing House of the Slovak Academy of Sciences, Bratislava, Czechoslovakia, pp. 312–317.

53. Feng, H.-M., Wen, J., and Walker, D. H., 1993, *Rickettsia australis* infection: A murine model of a highly invasive vasculopathic rickettsiosis, *Am. J. Pathol.* **142:**1471–1482.

54. Kern, W. V., Oristrell, J., Seguraporta, F., and Kern, P., 1996, Release of soluble tumor-necrosis-factor receptors in Mediterranean spotted fever rickettsiosis, *Clin. Diagn. Lab. Immunol.* **3:**233–235.

55. Nacy, C. A., Leonard, E. J., and Meltzer, M. S., 1981, Macrophages in resistance to rickettsial infections: Characterization of lymphokines that induce rickettsiacidal activity in macrophages, *J. Immunol.* **126:**204–207.

56. Nacy, C. A., and Groves, M. G., 1981, Macrophages in resistance to rickettsial infections: Early host defense mechanisms in experimental scrub typhus, *Infect. Immun.* **31:**1239–1250.

57. Nacy, C. A., and Osterman, J. V., 1979, Host defenses in experimental scrub typhus: Role of normal and activated macrophages, *Infect. Immun.* **26:**744–750.

58. Nacy, C. A., and Meltzer, M. S., 1979, Macrophages in resistance to rickettsial infection: Macrophage activation in vitro for killing of *Rickettsia tsutsugamushi*, *J. Immunol.* **123:**2544–2549.

59. Turco, J., and Winkler, H. H., 1983, Inhibition of the growth of *Rickettsia prowazekii* in cultured fibroblasts by lymphokines, *J. Exp. Med.* **157:**974–986.

60. Nacy, C. A., and Meltzer, M. S., 1982, Macrophages in resistance to rickettsial infection: Strains of mice susceptible to the lethal effects of *Rickettsia akari* show defective macrophage rickettsicidal activity in vitro, *Infect. Immun.* **36:**1096–1101.

61. Kodama, K., Yasukawa, M., and Kobayashi, Y., 1988, Effect of rickettsial antigen-specific T cell line on the interaction of *Rickettsia tsutsugamushi* with macrophages, *Microbiol. Immunol.* **32:**435–439.

62. Turco, J., and Winkler, H. H., 1985, Lymphokines, interferon-γ and *Rickettsia prowazekii*, in: *Rickettsiae and Rickettsial Diseases, Proceedings of the Third International Symposium* (J. Kazar, ed.), Publishing House of the Slovak Academy of Sciences, Bratislava, Czechoslovakia, pp. 211–218.

63. Turco, J., Liu, H., Gottlieb, S. F., and Winkler, H. H., Manuscript in preparation.

64. Liu, H., 1995, Effect of nitric oxide on the viability of *Rickettsia prowazekii* as measured by the ability to infect L929 cells and RAW264.7 cells, to grow in L929 cells and to kill gamma interferon-treated RAW264.7 cells, M.S. thesis, University of South Alabama, Mobile.

4

Global Genetic Diversity of Spotted Fever Group Rickettsiae

VERONIQUE ROUX and DIDIER RAOULT

1. INTRODUCTION

Members of the genus *Rickettsia* belong to the order *Rickettsiales*, a taxon of obligate intracellular symbionts, typically characterized as gram-negative short rods, which retain basic fuchsin when stained by the method of Gimenez.[1] Recent phylogenetic analysis of *Rickettsiales*, based on comparison of 16 S rRNA gene sequence, revealed a significant degree of evolutionary divergence within the order. The *Rickettsia* species were shown to lie within the α1 subgroup of the *proteobacteria*[2]; however, one member of the genus, *R. tsutsugamushi*, was considered different enough to be removed from the genus *Rickettsia* and was renamed *Orientia tsutsugamushi*.[3] The genus *Rickettsia* is therefore currently divided into two groups: the typhus group (TG), which includes three species (*R. prowazekii*, the agent of the epidemic typhus; *R. typhi*, which causes murine typhus; and *R. canada* isolated only from ticks), and the spotted fever group (SFG), which now includes an increasing number of serotypes following the development of a new cell culture isolation technique (the shell vial technique)[4] and a more widespread interest in arthropod-transmitted diseases (Table I). The SFG rickettsiae grow both in the nucleus and in the cytoplasm of their host cells. They are transmitted to humans by infected-arthropod bites or feces. The traditional identification methods used in bacteriology cannot be applied to rickettsiae because of the strictly intracellular

VERONIQUE ROUX and DIDIER RAOULT • Unité des rickettsies, Centre National de la Recherche Scientifique EP J 0054, Faculté de Médecine, 13385 Marseilles, France.

Rickettsial Infection and Immunity, edited by Anderson *et al.* Plenum Press, New York, 1997.

TABLE I
The Spotted Fever Group Rickettsiae

Rickettsiae	Arthropod vector	Geographic origin	Human disease
R. conorii	*Rhipicephalus sanguineus, R. simus, R. mushamae, H. leachi*	Morocco, South Africa, Ex USSR, India, Kenya	Mediterranean spotted fever
Astrakhan fever rickettsia	*Rhipicephalus pumilio*	Astrakhan region, Ex USSR	Astrakhan fever
Israeli TT rickettsia	*Rhipicephalus sanguineus*	Israel	Israeli spotted fever
R. sibirica	*Dermacentor nuttali, D. marginatus*	Ex USSR, China	Siberian tick typhus
"*R. mongolotimonae*" (HA-91)	*Haemophysalis asiaticum*	China, France	Unnamed disease[43]
R. parkeri	*Amblyomma maculatum*	Mississippi	
R. africae	*Amblyomma variegatum, A. hebraeum*	Shulu Province, Ethiopia, Zimbabwe	African tick bite fever
Strain S	*Rhipicephalus sanguineus*	Armenia, Ex USSR	
"*R. slovaca*"	*Dermacentor marginatus*	Slovakia, Switzerland, France, Crimea, Armenia	Unnamed disease[44]
Thai TT rickettsia	*Ixodes* or *Rhipicephalus* sp.	Thailand	
R. rickettsii	*Dermacentor andersoni, D. variabilis*	USA	Rocky Mountain spotted fever
R. japonica	*Haemophysalis longicornis*	Japan	Oriental spotted fever
R. massiliae	*Rhipicephalus sanguineus, R. turanicus*	France, Greece, Portugal	
Bar 29	*Rhipicephalus sanguineus*	Spain, France	
R. rhipicephali	*Rhipicephalus sanguineus*	Mississippi	
R. montana	*Dermacentor variabilis, D. andersoni*	Ohio	
"*R. aesch Pimannii*"	*Hyaloma marginatum*	Morocco	
R. australis	*Ixodes holocyclus*	Australia	Queensland tick typhus
R. akari	*Allodermanyssus sanguineus*	New York, Ukraine, Slovenia	Rickettsialpox
R. helvetica	*Ixodes ricinus*	Switzerland, France	
R. bellii	*Dermacentor variabilis*	USA	
R. felis	*Ctenocephalides felis*	Texas, California	Californian pseudotyphus
"*R. amblyommii*"	*Amblyomma americanum*	Missouri	
R. honei	?	Flinders Island, Tasmania	Flinders Island spotted fever
AB bacterium	*Adalia bipunctata*	England	

position of these bacteria and until recently their characterization has been based solely on morphological, pathogenic, ecological, and antigenic criteria. At present, serological typing by microimmunofluorescence (MIF) with mouse polyclonal antisera remains the reference method for the characterization of SFG rickettsiae, and the identification of a new species is based on values of "specificity difference" calculated from MIF results.[5] The antigenic determinants for this serotyping scheme are two high-molecular-weight outer membrane proteins, rOmpA and rOmpB. Newly recognized serotypes have also been characterized using other phenotypic criteria (protein analysis by SDS-PAGE and Western blot[6]); however, the introduction of molecular techniques has allowed new approaches to study the rickettsial diversity. With this aim, several studies have been performed on entire genomic DNA (DNA–DNA hybridization,[7] macrorestriction analysis,[8] and DNA restriction fragment polymorphism analysis[9,10]) or on a small part of the bacterial chromosome by restriction fragment length polymorphism analysis of PCR amplification products[11,12] or nucleotide base sequence comparison.[13–16]

2. STUDIES ON TOTAL DNA

2.1. DNA–DNA Hybridization

The widely accepted criterion for the definition of bacterial species is DNA–DNA hybridization. It has been proposed that two strains belonging to the same species share greater than 70% relatedness by DNA–DNA hybridization.[17] Few data are known for SFG rickettsiae as the purification of large amounts of DNA, which is prerequisite to this work, is very difficult and arduous for these fastidious bacteria. DNA relatedness as determined by DNA–DNA hybridization between *R. rickettsii* and other species is close: 91 to 94% for *R. conorii*, 70 to 74% for *R. sibirica*, 73% for *R. montana*, 53% for *R. australis*, and 46% for *R. akari*.[7] From these data, *R. rickettsii*, *R. conorii*, *R. sibirica*, and *R. montana* should belong to the same species, whereas *R. rickettsii* and *R. akari*, and *R. rickettsii* and *R. australis* should be considered as different species.

2.2. Macrorestriction Analysis by Pulsed Field Gel Electrophoresis

Determination of the Genome Size

The genome size has been determined for 16 SFG rickettsiae (*R. rickettsii*, *R. conorii*, Israeli tick typhus rickettsia, Thai tick typhus rickettsia, *R. africae*, *R. parkeri*, *R. sibirica*, "*R. slovaca*," *R. japonica*, *R. montana*, *R. rhipicephali*, *R. massiliae*, *R. australis*, *R. akari*, *R. helvetica*, and *R. bellii*) by pulsed field gel electrophoresis (PFGE).[8] If we exclude *R. bellii* for which phylogenetic position is unclear (it seems that this bacterium belongs neither to the typhus group nor to the spotted fever group), the size of the genome was between 1.2 and 1.3 Mb for all of the SFG rickettsiae tested except *R. helvetica* and *R. massiliae* (1.4 Mb).

It appears that the rickettsial chromosomes have not been subjected to major alterations (insertion or deletion of DNA) in the course of their evolution. In view of the phylogenetic studies inferred, we can propose that *R. massiliae* and *R. helvetica* genomes contain insert fragments of new DNA of approximately 150 kb.

Comparison of the PFGE Profiles

Three endonucleases were chosen to yield a suitable number of fragments for the comparison of profiles of the different rickettsiae of the SFG: *Eag*I, *Sma*I, and *Bss*HII.[8] It was possible to obtain different patterns with all of the rickettsiae, indicating that PFGE is a sensitive method to identify rickettsiae. The only limitation of the method is the requirement of a large quantity of DNA. The percentages of divergence between serotypes were estimated from profile differences using the dice coefficient. Most of the values were between 50 and 70%, although the divergence between *R. conorii* and Israeli tick typhus rickettsia was low (28%): this is not surprising as these two isolates belong to the *R. conorii* complex. Identical profiles were found for 9 isolates of *R. conorii*,[8] 7 isolates of *R. akari*,[18] and 15 isolates of *R. sibirica*[19]; however, differences were found between two isolates of *R. massiliae* (GS and Mtu1).[20]

2.3. DNA Restriction Fragment Polymorphism Analysis

Ralph et al. (1990)[9]

Rickettsial DNAs digested with the restriction enzymes *Rsa*I or *Hin*dIII were electrophoresed, blotted to nitrocellulose, and DNA probed with probes from *Eco*RI DNA fragments of *R. rickettsii* or *R. rhipicephali*. The method of Engles was chosen to analyze restriction fragment length polymorphism (RFLP) between serogroups and to estimate genetic similarity based on percent base-pair mismatch. Twelve serotypes were studied: *R. rickettsii*, *R. australis*, *R. akari*, *R. bellii*, *R. conorii*, *R. montana*, *R. parkeri*, *R. rhipicephali*, *R. sibirica*, "*R. slovaca*," *R. africae*, and Israeli tick typhus rickettsia.

The mean estimated difference between five *R. rickettsii* isolates examined was 1.7% and between two *R. conorii* isolates 1.2%, two *R. akari* isolates 1%, and two *R. bellii* isolates 1.3%. The values between *R. rickettsii* and *R. parkeri*, *R. sibirica*, "*R. slovaca*," *R. africae*, and *R. conorii* were 0.74, 1.15, 1.95, 1.7, and 2.35%, respectively. Thus, different species of rickettsia were found to be as closely related to *R. rickettsii* as *R. rickettsii* strains were to each other. *R. bellii*, *R. akari*, and *R. australis* were the most divergent species when compared to the other species examined but with percentages not above 17%.

Fuerst et al. (1990)[10]

Genetic homology between rickettsial strains was also compared using a combination of between 9 and 12 different random DNA probes and three to five different restriction enzymes. In total, the authors estimated the studied regions represented about 50–70 kb of DNA (3–5% of the total genome). *R. bellii, R. montana, R. rickettsii, R. rhipicephali,* and *R. sibirica* were included in this study. The genetic variability is expressed in terms of nucleotide diversity, estimated following the criteria of Nei and Li, equivalent to the average pairwise nucleotide difference between strains. Average levels of nucleotide diversity are extremely low. SFG interspecies variability is between 1.1 and 4.9% nucleotide differences and within *R. bellii, R. montana, R. rickettsii, R. rhipicephali,* and *R. sibirica* species it is 0.061–0.25, 0, 0.01, 0.49, and 0.23%, respectively.

3. RFLP ANALYSIS OF PCR AMPLIFICATION PRODUCTS AND BASE SEQUENCE COMPARISON OF PCR AMPLIFICATION PRODUCTS

Few genes have been sequenced in rickettsial genomes; however, five have now been used to study phylogenic organization or genomic diversity of *Rickettsia* species.

3.1. The Gene Coding for 16 S rRNA

The first gene to be sequenced in the rickettsial genome was the 1500-bp 16 S rRNA gene. Comparison of this gene allowed phylogenetic analysis of the rickettsia.[13,16] As observed among other bacteria, this gene is highly conserved among different representatives of the same genus. For rickettsiae, although a specific sequence was found for each serotype, extremely high sequence similarity between 99.9 and 97.2% was observed.

3.2. The Gene Coding for the Citrate Synthase (*glt* A)

Citrate synthase is a component of nearly all living cells and is an enzyme of the central metabolic pathway–citric acid cycle. *glt* A was the first cloned rickettsial gene.[21] By PCR amplification, the gene was shown to be present on the chromosome of all of the rickettsiae. Determination of PCR-RFLP profiles obtained after digestion of PCR products with *Alu*I was the first approach used to study phylogeny and genomic diversity of rickettsiae.[11,12] Only five serotypes showed characteristic profiles: *R. akari, R. australis, R. bellii, R. japonica,* and *R. massiliae.* All of the other studied SFG rickettsiae showed the same profiles. Roux *et al.* recently determined the *glt* A sequences for 24 rickettsiae.[15] The percent of

DNA similarity between *glt* A sequences was between 98.5 and 99.9% for most of the rickettsiae except: *R. helvetica, R. akari,* and *R. australis* (93.7 to 97.2% compared to other rickettsiae); *R. typhi, R. prowazekii,* AB bacterium, and *R. canada* (89.2 to 96.5%); *R. bellii* (85.4 to 95.8%).

3.3. The Genes Coding for the Proteins of 17 kDa, rOmpB, and rOmpA

The surface protein of 17 kDa is a genus-common antigen found on both the spotted fever and the typhus group rickettsiae.[22] Ten 17-kDa protein-encoding gene sequences were determined: those of *R. conorii, R. rickettsii, "R. amblyommii," R. japonica, R. australis, R. montana, R. parkeri,* and *R. rhipicephali* (P. A. Fuerst, unpublished data; Ref. 23). Within the SFG the observed similarities for the 17-kDa antigen gene were between 99.8% (*R. conorii/R. rickettsii*) and 95.1% (*R. australis/R. japonica*). The protein sequence obtained for *R. conorii* was 100% similar to that of *R. rickettsii.* The observed DNA similarities for the SFG representatives and the TG members were between 89.1% (*R. rickettsii/R. typhi*) and 87.0% (*R. australis/R. prowazekii*).

rOmpB and rOmpA are two high-molecular-weight outer membrane proteins. rOmpB is a surface-layer protein specific to the genus *Rickettsia* and rOmpA is a surface protein characterized on *R. canada* and most of the SFG rickettsial genomes by proteinic and genotypic analysis.[24] Presently, no biological function has been determined for these proteins, but their significance is evidenced by their capacity to be protective immunogens[25] and to react with convalescent immune sera.[26]

As discussed above, the serological reference method for the identification of rickettsiae exploits antigenic variation in the rOmpB and rOmpA proteins.

The gene encoding rOmpB, *omp* B, is 4912 bp long in the *R. rickettsii* genome[27] and 4836 bp long in the *R. prowazekii* genome.[28] Only these two complete sequences are presently available. Alignment of the *R. rickettsii* rOmpB gene with that of *R. prowazekii* revealed 78% nucleotide sequence similarity. Using primer pair BG1-21 and BG2-20 generated from *omp* B, for PCR amplification, and enzymatic digestion with *Rsa*I, Eremeeva *et al.* showed that most of the SFG rickettsiae had a specific profile, although *R. conorii* and Israeli tick typhus rickettsia, and *R. sibirica, R. parkeri,* and *R. africae*[12] could not be distinguished. Nucleotide differences in the 5' portion of the gene were described in different isolates of *R. rickettsii;* these differences corresponded to the segregation of virulent and avirulent strains.[27]

The gene encoding the protein rOmpA, *omp* A, is 6747 bp long in the *R. rickettsii* genome[29] and 6063 bp long in the *R. conorii* genome (Malish 7 strain).[30] Only these two complete sequences are presently available. Alignment of the *R.*

conorii rOmpA gene with that of *R. rickettsii* revealed 95% nucleotide sequence similarity.

The genetic structure of *omp* A can be considered to be made up of three parts: The structure of the *omp* A gene is characterized by a 700-bp region at the 5′ end of the gene separated from a 3254-bp region at the 3′ end by a series of near-identical repeating units of 216–225 bp. The 700-bp region at the 5′ end of the gene has been the focus of study on *omp* A, and variation within has been exploited for PCR-RFLP-based speciation schemes.

Using primers Rr 190.70p and Rr 190.602n for PCR amplification and enzymatic digestion with *Pst*I and *Rsa*I, Regnery *et al.*[11] and then Eremeeva *et al.*[12] were able to differentiate all SFG serotypes with the exception of *R. africae* and *R. parkeri*. Recently, Roux *et al.* sequenced this fragment for most of the described rickettsiae.[14] The degree of sequence similarity was between 99.3 and 91.2%. A different sequence was found for four isolates of the species *R. conorii* (Indian tick typhus rickettsia, M1, Seven, and Moroccan) but the same sequence was found for two isolates of the species *R. massiliae* (Mtu1 and GS).

The central region of *omp* A was sequenced for four rickettsiae: *R. conorii* (Kenya tick typhus rickettsia and Seven strain), *R. rickettsii*, and *R. akari*.[31] Differences were found in the number of repeating units and in their organization, even when comparing the two *R. conorii* isolates. By comparison of PCR product derived from this central region for a number of SFG rickettsiae (*R. sibirica, R. montana, R. rhipicephali, R. parkeri, R. africae*, Israeli tick typhus rickettsia, Astrakhan fever rickettsia),[24,32] variability in the size of the region resulting from a different number of repeated units was suspected. PCR/RFLP analysis and Southern blotting also identified differences in the internal structure of the repeat regions in different isolates of the species *R. rickettsii*.[24]

The introduction of molecular techniques has revolutionized the study of gene and genome evolution and has allowed new approaches to phylogenetic inference. Recently, a universal phylogenetic study of a vast spectrum of organisms has been achieved by comparison of 16 S rDNA sequences.[33] The order *Rickettsiales* has now been shown to be phylogeneticaly diverse and the genera *Coxiella* and *Rochalimaea* have been removed from the order.[2,34] The genera *Ehrlichia* and *Rickettsia* remained within the order *Rickettsiales* and although they share evolutionary homology, members of the genus *Ehrlichia* have been demonstrated to form three clearly distinct lineages[35] and *R. tsutsugamushi* has been demonstrated to form a distinct lineage from other members of this genus, resulting in a proposal for its transfer into a new genus, *Orientia*.[3] Presently, the precise organization within the genus *Rickettsia* is unclear and the phylogenetic studies undertaken have showed that a simple evolutionary division of species into either the TG or the SFG does not occur as at least three serotypes (*R. canada, R. bellii*, and AB bacterium) lie on a branch that diverged before the separation of these

TABLE II
Percentages of Similarity from *R. rickettsii* Data

Method	*R. conorii*	*R. sibirica*	*R. montana*	*R. akari*
Pulsed field gel electrophoresis	45	58	42	40
DNA–DNA homology	91–94	70–74	73	46
Fuerst	95.1–98.9	95.1–98.9	95.1–98.9	95.1–98.9
Ralph	97.65	98.85	97	83.3–94.5
Sequencing *glt* A	99.1	99.3	NT[a]	94
Sequencing 16 S rDNA	99.7	99.4	99.2	98.1
Sequencing *omp* A	94.6	95.9	90.7	NT
Sequencing 17-kDA encoding gene	99.8	NT	98.2	NT

[a]NT, not tested.

two groups.[15,36] The position of other recently characterized SFG rickettsiae is also unclear (*R. felis*,[37] *R. honei*,[38] AB bacterium,[39] "*R. amblyommii*"[40]) because we need more data concerning these bacteria. Indeed, the SFG itself may have no phylogenetic basis. At present, most of the genetic studies on the SFG examine only a few serotypes, and thus it is difficult to obtain an overview.

All of the approaches described above (DNA–DNA hybridization, PFGE, DNA restriction fragment polymorphism analysis, sequencing) involve estimation of genetic variation by detecting nucleotide differences in either specific regions of the genome or throughout the genome as a whole. Data from eight estimation methods are available for five *Rickettsia* SFG species (Table II). Comparison of these results shows some discrepancies in estimates of variability between different species (some of these differences are likely to have arisen from errors related to the method chosen or to statistical treatment of data), although some agreement between methods is also revealed. Variabilities between *R. rickettsii*, *R. conorii*, *R. sibirica*, and *R. montana* are generally very low, and thus it is unlikely that the differences observed between pairwise comparisons are significant. It is probably wise to conclude only that these four species are genotypically very similar. All data for *R. akari*, however, concur that this species is far less similar to *R. rickettsii* than the other three *Rickettsia* species studied. These results are also in agreement with phylogenetic inferences derived from either 16 S rRNA gene or *glt* A comparison, in which *R. akari* along with *R. australis* form the deepest diverging lineage within the SFG assembly.[13,15,16]

Interspecies similarity is very high as the evolutionary separation of these bacteria was very recent. If we apply the calibration of rRNA gene evolution proposed by Ochman and Wilson,[41] which holds that a 1% divergence between 16 S rRNA gene sequences would occur during 50 million years of divergence from a common ancestor, the divergence of TG and SFG occurred 120 million years ago and so the level of differentiation within the SFG is less than 100 million years old. Indeed, Fuerst *et al.* concluded that the interspecies diversity within the SFG (1.1 to 4.9%) is equivalent to that observed between different isolates of *E. coli* collected in the same locality (3–4%).[10]

The most difficult taxonomic problem for SFG rickettsiae is the definition of the species. At present, bacterial species demarcation is based on specific level of DNA–DNA hybridization.[17] Definition of rickettsial species using this criterion is, however, problematic. Interspecies DNA–DNA homology assessments were first determined among the Enterobacteriaceae, and the critical value of 70% was proposed based on the results relating to representatives of this family. The genome size of the *Rickettsia* species is four times smaller than those of enteric bacteria (1.0–1.4 Mb for most of the rickettsiae and 4.8 Mb for *E. coli*). It is hypothesized that as bacteria adapt from a free-living to an obligately intracellular lifestyle, bacterial genes that became unnecessary were discarded.[42] If this is the case, it is likely that the composition of the genome of these bacterial species existing in very similar niches is more similar. The observations established for free-living Enterobacteriaceae may therefore not necessarily be applicable to such scenarios.

Intrarickettsial similarity values are high regardless of the method of estimation employed, whether they are based on total DNA analysis or partial genome comparisons such as those of *omp* B and especially *omp* A. The detection of significant variation between strains of the same species is difficult, although preliminary analysis of the tandem repeat region of *omp* A suggests that a significant degree of variation may exist. More data are clearly necessary to confirm these preliminary observations. Examination of both intraspecific and interspecific similarity data strongly suggests that very little genetic transfer occurs between the rickettsial population. The obligate intracellular nature of the members of the genus *Rickettsia* is an evolutionary factor that sets them apart from free-living bacteria and the rickettsiae are also dependent for their continued survival on factors affecting the host. The transstadial–transovarial transfer of *Rickettsia* species in arthropod host does not favor genetic variability because contact with other bacteria is limited to other microorganisms transmitted by these arthropods (e.g., *Ehrlichia, Borrelia, Coxiella*). Thus, the opportunity for genetic recombination is low relative to the opportunity for genetic exchange between bacterial strains subjected to horizontal transfer between hosts and possible multiple infections.

Moreover the small size of the rickettsial chromosome indicates that only genes strictly necessary for their survival have been conserved, further limiting the possibility of recombination within the rickettsial genome.

The study of different rickettsial genes indicates that the speed of evolution within the genome is variable as dictated by the functionality of the specific gene product. This phenomenon is exploited for rickettsial identification schemes. Citrate synthase analysis allows characterization of all of the rickettsiae and the bartonellae too but not differentiation within the SFG. The rate of evolutionary change within the gene encoding rOmpA is significantly greater than that of *glt* A, and so its comparison is more useful for determining intra-SFG phylogeny. However, a precise picture of the genetic variability of the SFG rickettsiae can only be derived from comparison of an increasing number of genes which first require identification, then assessment.

ACKNOWLEDGMENT. We thank Richard Birtles for reviewing the manuscript.

REFERENCES

1. Weiss, E., and Moulder, J. W., 1984, Order I. *Rickettsiales* Gieszczkiewicz 1939, 25[AL], in: *Bergey's Manual of Systematic Bacteriology,* Volume 1 (N. R. Kreig and J. G. Holt, eds.), Williams & Wilkins, Baltimore, pp. 687–701.
2. Weisburg, W. G., Dobson, M. E., Samuel, J. E., Dasch, G. A., Mallavia, L. P., Baca, O., Mandelco, L., Sechrest, J. E., Weiss, E., and Woese, C. R., 1989, Phylogenetic diversity of rickettsiae, *J. Bacteriol.* **171:**4202–4206.
3. Tamura, A., Ohashi, N., Urakami, H., and Miyamura, S., 1995, Classification of *Rickettsia tsutsugamushi* in a new genus, *Orientia* gen. nov., as *Orientia tsutsugamushi* comb. nov., *Int. J. Syst. Bacteriol.* **45:**589–591.
4. Marrero, M., and Raoult, D., 1989, Centrifugation shell vial technique for rapid detection of Mediterranean spotted fever in blood culture. *Am. J. Trop. Med. Hyg.* **40:**197–199.
5. Philip, R. N., Casper, E. A., Burgdorfer, W., Gerloff, R. K., Hughes, L. E., and Bell, E. J., 1978, Serologic typing of *Rickettsiae* of spotted fever group by microimmunofluorescence, *J. Immunol.* **121:**1961–1968.
6. Anacker, R. L., Mann, R. E., and Gonzales, C., 1987, Reactivity of monoclonal antibodies of *Rickettsia rickettsii* with the spotted fever and typhus group rickettsiae, *J. Clin. Microbiol.* **25:**167–171.
7. Walker, D. H., 1989, Rocky Mountain spotted fever: A disease in need of microbiological concern, *Clin. Microbiol. Rev.* **2:**227–240.
8. Roux, V., and Raoult, D., 1993, Genotypic identification and phylogenetic analysis of the spotted fever group rickettsiae by pulsed-field gel electrophoresis, *J. Bacteriol.* **175:**4895–4904.
9. Ralph, D., Pretzman, C., Daugherty, N., and Poetter, K., 1990, Genetic relationships among the members of the family *Rickettsiaceae* as shown by DNA restriction fragment polymorphism analysis, *Ann. N.Y. Acad. Sci.* **590:**541–553.
10. Fuerst, P. A., Poetter, K. P., Pretzman, C., and Perlman, P. S., 1990, Molecular genetics of populations of intracellular bacteria: The spotted fever group rickettsiae, *Ann. N.Y. Acad. Sci.* **590:**430–438.
11. Regnery, R. L., Spruill, C. L., and Plikaytis, B. D., 1991, Genotypic identification of rickettsiae and estimation of intraspecies sequence divergence for portions of two rickettsial genes, *J. Bacteriol.* **173:**1576–1589.
12. Eremeeva, M., Y, X., and Raoult, D., 1994, Differentiation among spotted fever group rickettsiae species by analysis of restriction fragment length polymorphism of PCR-amplified DNA, *J. Clin. Microbiol.* **32:**803–810.
13. Roux, V., and Raoult, D., 1995, Phylogenetic analysis of the genus *Rickettsia* by 16S rDNA sequencing, *Res. Microbiol.* **146:**385–396.

14. Roux, V., Fournier, P. E., and Raoult, D., 1996, Differentiation of spotted fever group rickettsiae by sequencing and analysis of restriction fragment length polymorphism of PCR amplified DNA of the gene encoding the protein rOmpA, *J. Clin. Microbiol.* **34**:2058–2065.

15. Roux, V., Rydkina, E., Eremeeva, M., and Raoult, D., 1997, Citrate synthase gene comparison, a new tool for phylogenetic analysis, and its application for the rickettsiae, *Int. J. Syst. Bacteriol.* **47**:252–261.

16. Stothard, D. R., and Fuerst, P. A., 1995, Evolutionary analysis of the spotted fever and typhus groups of *Rickettsia* using 16S rRNA gene sequences, *Syst. Appl. Microbiol.* **18**:52–61.

17. Wayne, L. G., Brenner, D. J., Colwell, R. R., Grimont, P. A. D., Kandler, O., Krichevsky, M. I., Moore, L. H., Moore, W. E. C., Murray, R. E. G., Stackebrandt, E. P., Starr, M., and Trüper, H. G., 1987, Report of the ad hoc committee on reconciliation of approaches to bacterial systematics, *Int. J. Syst. Bacteriol.* **37**:463–464.

18. Eremeeva, M., Balayeva, N., Ignatovich, V., and Raoult, D., 1995, Genomic study of *Rickettsia akari* by pulsed-field gel electrophoresis, *J. Clin. Microbiol.* **33**:3022–3024.

19. Eremeeva, M., Balayeva, N. M., Ignatovich, V. F., and Raoult, D., 1993, Proteinic and genomic identification of spotted fever group rickettsiae isolated in the former USSR, *J. Clin. Microbiol.* **31**:2625–2633.

20. Babalis, T., Tselentis, Y., Roux, V., Psaroulaki, A., and Raoult, D., 1994, Isolation and identification of a rickettsial strain related to *Rickettsia massiliae* in Greek ticks, *Am. J. Trop. Med. Hyg.* **50(3)**:365–372.

21. Wood, D. O., Williamson, L. R., Winkler, H. H., and Krause, D. C., 1987, Nucleotide sequence of the *Rickettsia prowazekii* citrate synthase gene, *J. Bacteriol.* **169**:3564–3572.

22. Anderson, B. E., Regnery, R. L., Carlone, G. M., Tzianabos, T., McDade, J. E., Fu, Z. Y., and Bellini, W. J., 1987, Sequence analysis of the 17-kilodalton-antigen gene from *Rickettsia rickettsii*, *J. Bacteriol.* **169**:2385–2390.

23. Anderson, B. E., and Tzianabos, T., 1989, Comparative sequence analysis of a genus-common rickettsial antigen gene, *J. Bacteriol.* **171**:5199–5201.

24. Gilmore, R. D., Jr., and Hackstadt, T., 1991, DNA polymorphism in the conserved 190 kDa antigen gene repeat region among spotted fever group rickettsiae, *Biochim. Biophys. Acta* **1097**:77–80.

25. McDonald, G. A., Anacker, R. L., Mann, R. E., and Milch, L. J., 1988, Protection of guinea pigs from experimental Rocky Mountain spotted fever with a cloned antigen of *Rickettsia rickettsii*, *J. Infect. Dis.* **158**:228–231.

26. Anacker, R. L., List, R. H., Mann, R. E., Hayes, S. F., and Thomas, L. A., 1985, Characterization of monoclonal antibodies protecting mice against *Rickettsia rickettsii*, *J. Infect. Dis.* **151**:1052–1060.

27. Gilmore, R. D., Jr., Cieplak, W., Jr., Policastro, P. F., and Hackstadt, T., 1991, The 120 kilodalton outer membrane protein (rOmpB) of *Rickettsia rickettsii* is encoded by an unusually long open reading frame: Evidence for a protein processing from a large precursor, *Mol. Microbiol.* **5(10)**:2361–2370.

28. Carl, M., Dobson, M. E., Ching, W.-M., and Dasch, G. A., 1990, Characterization of the gene encoding the protective paracrystalline-surface-layer protein of *Rickettsia prowazekii*: Presence of a truncated identical homolog in *Rickettsia typhi*, *Proc. Natl. Acad. Sci. USA* **87**:8237–8241.

29. Anderson, B. E., McDonald, G. A., Jones, D. C., and Regnery, R. L., 1990, A protective protein antigen of *Rickettsia rickettsii* has tandemly repeated, near-identical sequences, *Infect. Immun.* **58**:2760–2769.

30. Croquet-Valdes, P. A., Weiss, K., and Walker, D. H., 1994, Sequence analysis of the 190-kDa antigen-encoding gene of *Rickettsia conorii* (Malish 7 strain), *Gene* **140**:115–119.

31. Gilmore, R. D., Jr., 1993, Comparison of the *romp* A gene repeat regions of *Rickettsiae* reveals species-specific arrangements of individual repeating units, *Gene* **125**:97–102.

32. Walker, D. H., Feng, H.-M., Saada, J. I., Crocquet-Valdes, P., Radulovic, S., Popov, V. L., and Manor, E., 1995, Comparative antigenic analysis of the spotted fever group rickettsiae from israel and other closely related organisms, *Am. J. Trop. Med. Hyg.* **52(6):**569–576.

33. Woese, C. R., 1987, Bacterial evolution, *Microbiol. Rev.* **51:**221–271.

34. Brenner, D. J., O'Connor, S. P., Winkler, H. H., and Steigerwalt, A. G., 1993, Proposals to unify the genera *Bartonella* and *Rochalimaea*, with descriptions of *Bartonella quintana* comb. nov., *Bartonella vinsonii* comb. nov., *Bartonella henselae* comb. nov., and *Bartonella elizabethae* comb. nov., and to remove family *Bartonellaceae* from the order *Rickettsiales, Int. J. Syst. Bacteriol.* **43:**777–786.

35. Anderson, B. E., Dawson, J. E., Jones, D. C., and Wilson, K. H., 1991, *Ehrlichia chaffeensis, a new* species associated with human ehrlichiosis, *J. Clin. Microbiol.* **29:**2838–2842.

36. Stothard, D. R., Clark, J. B., and Fuerst, P. A., 1994, Ancestral divergence of *Rickettsia bellii* from the spotted fever and the typhus groups of *rickettsia* and antiquity of the genus *Rickettsia, Int. J. Syst. Bacteriol.* **44:**798–804.

37. Higgins, J. A., Radulovic, S., Schrieffer, M. E., and Azad, A. F., 1996, *Rickettsia felis:* A new species of pathogenic rickettsia isolated from cat fleas, *J. Clin. Microbiol.* **34:**671–674.

38. Baird, R. W., Lloyd, M., Stenos, J., Ross, B. C., Stewart, R. S., and Dwyer, B., 1992, Characterization and comparison of Australian human spotted fever group rickettsiae, *J. Clin. Microbiol.* **30:**2896–2902.

39. Werren, J. H., Hurst, G. D. D., Zhang, W., Breeuvwer, J. A. J., Stouthamer, R., and Majerus, M. E. N., 1994, Rickettsial relative associated with male killing in the ladybird beetle (*Adalia bipunctata*), *J. Bacteriol.* **176:**388–394.

40. Pretzman, C., Stothard, D. R., Ralph, D., and Fuerst, A., 1994, A new *Rickettsia*, isolated from the lone star tick, *Amblyomma americanum* (*Ixodidae*), Abstracts 11th Sesqui-annual Meeting of the American Society for Rickettsiology and Rickettsial Diseases, St. Simons Island, GA, p. 24.

41. Ochman, H., and Wilson, A. C., 1987, Evolution in bacteria: Evidence for a universal substitution rate in cellular genomes, *J. Mol. Evol.* **26:**74–86.

42. Weisburg, W. G., 1989, Polyphyletic origin of bacterial parasites, in: *Intracellular Parasitism* (J. W. Moulder, ed.), CRC Press, Boca Raton, pp. 1–15.

43. Raoult, D., Brouqui, P., and Roux, V., 1996, A new spotted fever group rickettsiosis, *Lancet* **348:**412.

5

Surface Components of the Spotted Fever Group Rickettsiae

GREGORY A. MCDONALD, CAYLE C. GRAUMANN, and JOSEPH J. TEMENAK

1. SURFACE ORGANIZATION OF THE SPOTTED FEVER GROUP RICKETTSIAE

The spotted fever group (SFG) of the genus *Rickettsia* is composed of several species that are related antigenically and genetically. Members of this group have been isolated from a large worldwide distribution (Table I). The two species most readily associated with human disease, and most extensively studied, are *R. rickettsii* and *R. conorii*, the etiological agents of Rocky Mountain spotted fever (RMSF) and boutonneuse fever, respectively. This chapter will focus on information available on these two species and strains therein. Special attention will be given to the inclusion of other SFG species when appropriate.

Although once thought to be viruses, SFG rickettsiae (and all members of this genus) have an envelope structure typical of gram-negative bacteria. As such, they have trilaminar inner and outer membranes, a periplasmic space, and a peptidoglycan layer between the membranes.[1,2] The diameter of the inner membrane is 5.7 nm and that of the outer membrane is 13.0 nm.[2] The reported diameter for the outer membrane most probably includes the diameter of the closely associated[2] peptidoglycan layer. Evidence strongly supports the presence

GREGORY A. MCDONALD, CAYLE C. GRAUMANN, and JOSEPH H. TEMENAK • Department of Molecular Microbiology and Immunology, School of Medicine, University of Missouri, Columbia, Missouri 65212.

Rickettsial Infection and Immunity, edited by Anderson *et al.* Plenum Press, New York, 1997.

TABLE I
SFG Species and Strains Discussed

Rickettsial species	Disease	Location	MAb to rOmpA	MAb to rOmpB
R. africae	Unknown	Africa		
R. akari	Rickettsialpox	North America, Asia		
R. australis	Queensland tick typhus	Australia		
R. bellii[a]	Avirulent	North America		
R. conorii[b]	Boutonneuse fever	Africa, India, Europe, Middle East	+	+
R. japonica	Oriental spotted fever	Japan, China	+	
R. montana	Avirulent	North America	+	+
R. parkeri	Avirulent		+	+
R. rhipicephali	Avirulent		+	
R. riskettsii	RMSF	North and South America	+	+
R. sibirica	North Asian tick typhus	Asia	+	
R. slovaca	Avirulent	Europe, Asia		
R. sp. TT-118[c]	Unknown	Asia		
R. sp. 363-D	Avirulent	North America	+	+

[a]Formerly known as *R.* sp. 369-C.
[b]The species designation *R. conorii* includes those species formerly designated *R. israeli, R. zimbabwe,* and the Astrakhan fever agent.
[c]*R.* sp. TT-118 is also known as the Thai tick typhus organism.

of lipopolysaccharide (LPS) in association with the outer membrane. Also, the total fatty acid composition of these organisms is similar to that of gram-negative bacteria.[3]

Two additional layers exist external to the outer membrane. The microcapsular layer (MCL), which immediately surrounds the outer membrane, is 16 nm thick.[2] Surrounding the microcapsular layer is the slime layer (SL)[1,2] which has a variable thickness of ≤ 125 nm.[2] Both of these surface components have polysaccharides associated with them and have been best described for the typhus group rickettsiae. More about the LPS, MCL, and SL will be presented later in the chapter.

2. PROTEIN SURFACE COMPONENTS

Different protein detection methods have been used to determine the total number of proteins associated with these organisms. These methods include SDS-PAGE followed by staining with Coomassie brilliant blue, and SDS-PAGE, coupled with autoradiography, on rickettsiae intrinsically labeled with radio-labeled amino acids. In one of the earliest investigations, Obijeski *et al.*[4] reported SDS-PAGE analyses performed on purified *R. rickettsii*, *R. akari*, and three members of the typhus group rickettsiae. A minimum of 30 proteins were reported for *R. rickettsii* and a minimum of 27 were reported for *R. akari*. After careful scrutiny of studies published from 1974 to 1978, Feng *et al.*[5] determined that a reasonable approximation of the number of "catalogued" proteins for *R. rickettsii* and *R. conorii* would be 35. On careful examination of protein profiles of *R. rickettsii* dually labeled with rabiolabeled methionine and leucine, a minimum of 50 peptides have been detected (McDonald and Graumann, unpublished observations). It is not clear, however, that all of these peptides represent different proteins. The true total number of rickettsial proteins awaits to be determined. Regardless of the total number of proteins, the questions are, how many and which of these proteins are associated with the surface?

2.1. Proteins Identified with Polyclonal Antibodies

SDS-PAGE analysis of components from extrinsically labeled (surface-iodinated) rickettsiae has provided some information in the search for surface proteins. Six proteins were thus identified on the surface of *R. conorii*; three of the six were also intrinsically labeled with [³H]galactose, implying that they are glycoproteins.[6] Convalescent sera from guinea pigs previously infected (21 days) with *R. rickettsii* were used in radioimmunoprecipitation assays on detergent extracts from five extrinsically labeled *R. rickettsii* strains.[7] Proteins with apparent molecular masses of 128, 105, 84, 30.5, and 20.5 kDa were detected. In a similar experiment, Williams *et al.*[8] detected proteins, from surface-iodinated *R. rickettsii*, with apparent molecular masses of 186, 145, 49, 32, 27.5, 17.5, and 16.5 kDa; these same sera, which were collected 163 days after infection of the guinea pigs, also reacted (as measured by immunodiffusion) with three soluble antigens (two proteins and one nonprotein) obtained from tissue culture supernatants of *R. rickettsii*-infected L cell monolayers. Convalescent human sera (from RMSF patients) also precipitated up to seven major proteins from a detergent extract of extrinsically labeled *R. rickettsii*. These proteins were within the range from 190 to 17 kDa.[9]

2.2. Surface Proteins Identified with Monoclonal Antibodies

Perhaps the greatest advancement toward the identification of specific SFG rickettsiae surface proteins came with the utilization of monoclonal antibody (MAb) technologies. The first to report the production of MAbs to a member of this group of organisms were Lange and Walker.[10] They reported results obtained with five murine MAbs raised to the *R. rickettsii* Sheila Smith strain. When tested in the indirect immunofluorescence assay (IFA), all five MAbs exhibited high titers to the homologous strain as well as two other *R. rickettsii* strains. These antibodies had reduced IFA titers to *R. montana* and no significant titers to *R. akari*, *R. sibirica*, and *R. conorii*. Also, all five MAbs were positive in microagglutination assays (MA) when purified *R. rickettsii* was used as the antigen. Potential biological relevance was ascribed to these five MAbs when they proved capable of passively protecting mice from "toxic" death in the mouse protection test (MPT).* Two of the five were also successful in passively preventing rickettsemia and death, as well as reducing morbidity in experimentally induced RMSF in the guinea pig model of infection.† The data rendered in the IFA, MA, and possibly the passive protection tests in the animal models, indicated that these MAbs recognized surface components. However, as well performed and reported as these studies were, there is no evidence that the MAbs bound to proteins and not to other surface components.

Further research using MAbs has led to a more definitive identification of surface proteins. In 1985, Anacker *et al.*[9] reported the characterization of 31 murine MAbs reactive with two strains (Sheila Smith and Morgan) of *R. rickettsii*. In radioimmunoprecipitation assays (a detergent extract of surface-iodinated rickettsiae served as antigen), 20 of the MAbs recognized a protein with an apparent molecular mass of 170 kDa and six MAbs precipitated polypeptides of 133 and 32 kDa. Some of these MAbs were also capable of recognizing surface-exposed epitopes when tested in immunoelectron microscopy studies on partially purified *R. rickettsii*. Thus, surface locations for these proteins were directly established; this conclusion was further supported by the demonstration that some MAbs were also positive in ELISA and microimmunofluorescence (MIF) assays performed on intact rickettsiae. Nineteen of the twenty MAbs to the 170-kDa protein, and all six to the 133-kDa protein, protected mice in the MPT, thereby suggesting a role for these proteins in *R. rickettsii* pathogenesis. The researchers were not able to demonstrate MAb reactivity to the 170- and 133-kDa proteins by Western immunoblots. This was, however, accomplished and reported in a later study by Anacker *et al.*[11] In the latter study, *R. rickettsii* organisms were solubilized in 4% SDS at a temperature ≤37°C for 15–120 min rather than 100°C. After solubilization at the lower temperatures and SDS-PAGE on 10% gels (instead of 12.5% gels), MAbs, which previously immunoprecipitated the "170-kDa" protein, reacted (in

---

*The mouse protection test was first described by Bell and Pickens.[44] It is based on the observation that immune sera from humans and guinea pigs passively protected mice from a "toxic" death which would normally occur within 24 hr after the mice received an intravenous injection of SFG rickettsiae.

†Infection of guinea pigs with virulent SFG rickettsiae results in disease with signs that are characteristic of RMSF in humans.[45] These include fever, weight loss, and skin necrosis. Li *et al.*[46] proposed that the mechanism by which MAbs passively protect guinea pigs and mice may be by opsonization and enhancement of phagocytosis or by blocking the ability of rickettsiae to adhere to and/or enter endothelial cells.

Western immunoblots) to a 155-kDa peptide. Those that previously immunoprecipitated the 133- and 32-kDa proteins reacted (in Western immunoblots) to a 120-kDa protein but not one of 32 kDa. These MAbs did not react with any proteins when the rickettsiae were solubilized at 100°C. At this point, it was established and well accepted that the two surface antigens recognized by Anacker's MAbs were the R. rickettsii "155- and 120-kDa proteins" previously reported to be 170 and 133 kDa, respectively.[9] As evidenced by ELISA assays, MAbs to the 155- and 120-kDa proteins reacted to the surfaces of several R. rickettsii strains (R, Sheila Smith, Morgan, Wachsmuth, Norgaard, Simpson, HLP, 275-F) and other SFG species including R. parkeri, R. conorii, R. montana, and Rickettsia 364-D. Interestingly, only the 155-kDa protein was detected on R. sibirica and R. rhipicephali, and neither protein was demonstrable on the surfaces of R. akari or R. australis.

A potential role of the 155- and 120-kDa proteins in the pathogenesis of R. rickettsii was further strengthened by the identification of more MAbs to these proteins which were capable of protecting mice (protective MAbs) in the MPT.[12] The protective MAbs to the 120- and 155-kDa proteins were shown to recognize heat-sensitive epitopes. This was demonstrated through the observation that if whole-cell R. rickettsii organisms were heated at 80°C and above, protective MAbs no longer reacted in dot-blot assays, while the reactivity of nonprotective MAbs increased. These data suggested that epitopes recognized by protective MAbs are conformational. One can easily envision that the regions of these proteins that constitute "conformational" epitopes may serve important structural and/or functional purposes. This, however, remains to be proven.

The existence of heat-sensitive, protective epitopes on surface proteins has been established for other SFG rickettsiae also. Feng et al.[13] were the first to describe the production of MAbs to R. conorii. Of 6 MAbs characterized, one (MAb 3-2) was demonstrated to be a protective MAb because it conferred 100% protection to mice in the MPT. This antibody recognized a heat-labile, surface-exposed epitope on a 112-kDa protein. As evidenced by the IFA test, this MAb also reacted with proteins on the surfaces of three strains of R. rickettsii, R. sibirica, and two nonpathogenic rickettsiae (TT-118 and R. slovaca), but not with R. akari. Li et al.[14] described results obtained with MAbs raised against R. rickettsii, R. conorii, and R. sibirica. Thirty-eight MAbs were evaluated; 13 were from mice immunized with R. rickettsii, 23 were from mice immunized with R. conorii, and 2 were derived from mice immunized with R. sibirica. Of the MAbs made to R. conorii, all recognized heat-sensitive epitopes on proteins of 135 or 115 kDa and passively protected mice in the MPT. As determined by immunoelectron microscopy, 22 of the R. conorii MAbs reacted with surface-exposed epitopes. In MIF assays, 6 of the R. conorii MAbs were species specific (including the one MAb that recognized the 115-kDa protein); 6 also reacted with four other species of SFG rickettsiae (R. rickettsii, R. sibirica, R. akari, Rickettsia strain TT-118); the remaining

11 reacted with two to four of the other SFG species. MAbs to the *R. conorii* 135-
and 115-kDa proteins were successful in passively protecting mice in the MPT
against challenge with *R. conorii*. The *R. conorii* strain used in this study was Malish
7. Walker *et al.*[15] later demonstrated that some of the species-specific MAbs
reacted with six other *R. conorii* strains while some exhibited selective reactivity
between the strains. For *R. rickettsii*, heat-sensitive surface-exposed epitopes were
demonstrated to 150- and 135-kDa proteins, while surface proteins bearing heat-
sensitive epitopes on *R. sibirica* were found on proteins of 130 and 115 kDa. Three
of the MAbs directed against the *R. rickettsii* proteins (two to the 150-kDa and one
to the 135-kDa protein) were capable of providing complete passive protection,
against homologous challenge, to guinea pigs in the guinea pig model of infection.
The two *R. sibirica* MAbs provided partial passive protection against homologous
challenge. It should be noted that some MAbs to these 150-, 135-, and 115-kDa
proteins were also broadly reactive (as evidenced by immunofluorescence) to
surface epitopes on *R. conorii* (150-, 135-, and 115-kDa proteins), *R. rickettsii* (150-,
135-, and 115-kDa proteins), *R. akari* (135-kDa protein only), *R. sibirica* (150-,
135-, and 115-kDa proteins), and *Rickettsia* strain TT-118 (150-, 135-, and 115-
kDa proteins).

SFG rickettsial infections in Japan were first documented during the late
1980s. *Rickettsia* spp. were isolated from patients by Uchida *et al.*[16,17] Mouse
antisera raised to five Japanese isolates exhibited positive, although reduced,
reactivity to *R. conorii*, *R. rickettsii*, *R. sibirica*, *R. akari*, *R. australis*, and *Rickettsia*
strain TT-118.[18] MAbs have been extremely useful in characterizing this species
which is now known as *R. japonica*. Although reactive with mouse antisera raised
against SFG rickettsiae, the organism reacted poorly or not at all with species-
specific MAbs to *R. conorii*, *R. rickettsii*, *R. sibirica*, or *R. akari*.[18] MAbs raised to *R.
japonica* reacted to heat-sensitive epitopes on surface proteins with apparent molecular masses of 120 and 145 kDa.[19] Close associations of these proteins to the
surface of this organism were demonstrated by immunoelectron microscopy on
ultrathin sections made from whole-cell bacteria.[20]

The studies mentioned above do not represent all of the investigations that
have utilized MAbs in the search for SFG surface proteins. They do, however,
include major findings that set the stage for a surge in molecular analyses of
surface proteins and the genes encoding them. A common theme among these
studies is the existence of at least two major surface proteins, on each of these
organisms, which have apparent molecular masses of >100 kDa.

2.3. Rickettsial Outer Membrane Proteins

Extensive research has been conducted on two SFG surface proteins now
called rOmpA and rOmpB. These proteins were first termed "rickettsial outer
membrane protein A (rOmpA)" and "rickettsial outer membrane protein B

(rOmpB)" by Gilmore *et al.*[21] It is well established that rOmpA and rOmpB are the same proteins as Anacker's *R. rickettsii* 155- and 120-kDa proteins, respectively; homologues of these proteins have been identified in other SFG rickettsiae and are also referred to as rOmpA and rOmpB. Although the "romp" terminology is widely accepted, it is a very curious designation as neither of these proteins has been shown to be an "outer membrane" protein, although they are clearly associated with the outer surface. So as to be consistent with the literature and not "muddy" the waters any further, this terminology will be used here.

rOmpA

Using molecular cloning techniques, McDonald *et al.*[22] reported the cloning and expression, in *Escherichia coli*, of a portion (4.1 kb) of the *R. rickettsii* 155-kDa protein *rompa* gene. The recombinant peptide reacted with MAbs to the 155-kDa protein and, when used as a vaccine, elicited protective immunity in mice against a potentially lethal injection of *R. rickettsii*. In a later study, McDonald *et al.*[23] demonstrated that the recombinant peptide successfully immunized guinea pigs from experimentally induced RMSF. This recombinant was also useful in establishing the identity of the *R. rickettsii* "antigen 2." Anacker *et al.*[24] had earlier demonstrated, through immunoelectron microscopy, that antigen 2 was located on the surface of *R. rickettsii* and that monospecific antibodies raised to this antigen passively protected mice and guinea pigs against potential disease on challenge with viable *R. rickettsii*. Samples of these same monospecific antibodies were reactive with the recombinant protein (McDonald, unpublished observations).

Anderson *et al.*[25] provided a momentous boost to the study of the *rompa* gene. They were successful in cloning and sequencing the gene in its entirety. The deduced molecular mass for the protein was reported to be 190 kDa. Consequently, the 155-kDa protein became known, for a short period, as the *R. rickettsii* 190-kDa protein. Sequence analysis of this gene revealed properties common to prokaryotic genes. The putative -35 promoter region (TAGACA) matches 5 of 6 bases with the *E. coli* consensus sequence and the putative -10 promoter sequence (TATAAC) showed a 5 of 6 match with the *E. coli* -10 consensus sequence. By use of primer extension analysis, it was directly determined that the transcriptional start $(+1)$ site for the cloned *rompa* promoter was the same as that used in *R. rickettsii*, and was located at 8 bases distal to the putative -10 site. This $+1$ site mapped at 19 bases proximal to the putative ribosome-binding site (RBS) which was 8 bases proximal to a presumed state methionine. Distal to the TAA stop codon is a 16-base inverted repeat capable of forming a stem and loop structure with a ΔG of -28.3 kcal/mol; such a structure is reminiscent of a prokaryotic transcriptional stop.

Within a 2861-bp portion of the gene (starting at nucleotide 703 and ending at nucleotide 3564) there exists a series of 13 highly conserved direct repeats, each encoding 75 amino acids (type I) or 72 amino acids (type II). The amino acid sequence of the repeats are shown in Table II. Repetitive regions have been reported for several bacterial proteins including the streptococcal M protein, the Vlp proteins of *Mycoplasma* spp., the *Neisserria gonorrhoeae* opacity proteins, and the *Anaplasma marginale* MSP-2 protein. However, the RompA repeats are the longest yet identified. The number of these repeats has been found to vary among SFG rickettsiae and has been extensively exploited for species/strain differentiation. For the sake of simplicity, the repeats will be considered as a single unit and will be referred to as the "RompA repeat cassette."

The sequence of the *R. conorii* (Malish 7 strain) *rompa* gene has been reported[26] (Table II). This gene is 95% homologous to that in *R. rickettsii* and this conservation extends to the putative promoter and transcription termination regions. The deduced amino acid sequence revealed a protein of 203.2 kDa with ten repeats (type I, 75 amino acids; type II, 72 amino acids) in the RompA repeat cassette (Table II).

As mentioned earlier, variations in the number of repeats exist among different SFG rickettsiae (Table II). Determination of differences in repeat numbers in the RompA repeat cassette, for species differentiation, was first directly investigated and reported by Gilmore,[27] who DNA sequenced the RompA repeat cassettes of *R. conorii* and *R. akari*. This analysis revealed that the *R. conorii* (KTT strain) protein has 14 repeats and the *R. akari* (Kaplan strain) protein contains 11. Walker *et al.*[28] estimated the number of repeats in the RompA cassette for the SFG Astrakhan isolate, four SFG isolates from Israel (*R. conorii*), the *R. conorii* Malish 7 and Moroccan strains, and *R. africae*. The number of repeats for the Israeli isolates and the Astrakhan strain were 15, the *R. conorii* Moroccan strain and *R. africae* had 6, and the *R. conorii* Malish 7 strain had 10.

No function has yet been identified for RompA, but the search continues. As participants in this endeavor, Li *et al.*[29] reported that a panel of five MAbs to this protein blocked the attachment of *R. rickettsii* to host cells. These results are indeed promising and may lead the way for further RompA studies.

rOmpB

Anacker's 120-kDa protein (now called rOmpB) has received less attention as has rOmpA, but important information is available. The first *rompb* sequence (from the *R. rickettsii* R strain) was molecularly cloned and expressed, in *E. coli*, and reported by Gilmore *et al.*[30] Further sequence analysis later revealed that the DNA sequence that Gilmore reported was lacking 814 bp from the 5′ region of the structural gene. The sequence of the remainder of the gene, in addition to 197 bp proximal to the translated region, was later reported by Gilmore *et al.*[21] Putative promoter −10 and −35 regions were identified, and a +1 mRNA start site was mapped to 129 bases proximal to the methionine start codon on the

TABLE II

Amino Acid Sequences of rOmpA Repeat Cassettes

Rickettsial species	Repeat unit	No.	Amino acid sequence[a]
R. akari[b]	Type I	9	IGN[TV]N [SP][LA]A[TQ]I SVGA[GS] [TP][AL][TS]LG GAVIK ATTTD LT[DN]AA SVLTL TNANA VLTG[AV] [IV]DNTT GVDNV GVLNL NGALS Q[VA]TGN
	Type II	2	IGNT[NA] [SA]LAT[IV] [SN]VGAG [TL][AL][TQ][LV][GQ] G[AG]V[IV]K A[NT][AT][IT][KN] LTD[AN][AV] A[AQ]VTF TNPVV VTGAI DNTGN ANNGI ATFTG NSTVT GN
R. conorii KTT strain[c]	Type I	2	IGNTN AL[RA]TV NVGAG IATLE GA[IV]K ATTTK LTNAA SVLTL TNVNA VLTGA IDNTT GVDNV GVLNL NGALS QVTGN
	Type II	12	IGNTN ALAT[IV] [SN]VGAG [KL][AL][RT][LV][GQ] G[AG][IV][IV]K [AS][NT][T][IT][KN] LTDNA SQVTF TNPVV VTGAI DNTGN ANNGI [VA]TFTG [DGN]STVT GN
R. conorii Malish 7 strain[d]	Type I	2	IGNTN ALATV NVGAG IATLE GAIIK ATTTK LTNAA SVLTL TNVNA VLTGA IDNTT GVDNV GVLNL NGALS QVTGN
	Type II	8	IGNTN ALAT[IV] [SN]VGAG [KL][AL][QT][LV][GQ] G[AG][VI][VI]K A[TN][T][TI][KN] LT[DT]NA SAVTF TN[NP]VV VTGAI DNTGN ANNG[IL] VTFTG DSTVT G[DN]
R. rickettsii[e]	Type I	6	IGNTN [AS]LAT[VI] [NS]VGAG TATLG GAVIK ATTTK LTNAA SVLTL TNANA VLTGA [IV]DNTT GGDNV GVLNL NGALS QVTG[DN]
	Type II	7	[IV]GNTN [AS]LAT[VI] [SN]VGAG [TL][AL][QT][VL][QG] G[AG]V[IV]K A[NT][T][IT][NK] [IL]T[DN][AN]A SAV[TK]F TNPVV VTGAI D[SN[TGN ANNGI VTFTG NSTVT G[DN]

[a] Amino acids within brackets are sequence variations.
[b] Reference 27.
[c] Kenyan tick typhus strain.[27]
[d] Reference 26.
[e] Reference 25.

cloned gene. This same +1 site is used for this gene in intact *R. rickettsii.*[31] The deduced molecular mass for rOmpB was reported to be 167.9 kDa, consisting of 1654 amino acids with a predicted N-terminal secretory signal sequence of 34 residues. The protein sequence was found to be 65% homologous to the rOmpB analogue (initially referred to as the surface protein antigen[32]) of the typhus group rickettsia, *R. prowazekii.* The sequence for the *R. prowazekii* surface protein antigen was determined by Carl *et al.*[33]; previous data[34] allowed for the proposal that the *R. prowazekii* SPA forms the S-layer surrounding this organism. It is not clear whether the *R. rickettsii* rOmpB also forms an S-layer.

Hackstadt *et al.*[35] later provided data demonstrating that the *R. rickettsii* rOmpB is made as a 168-kDa protein that is posttranslationally cleaved to yield 120- and 32-kDa peptides. The 32-kDa peptide was heat modifiable in SDS-PAGE analysis. Both peptides were demonstrated to be surface exposed as evidenced by extrinsic iodination, and both were radioimmunoprecipitated by MAbs to rOmpB. This coprecipitation led the authors to suggest that a close, but noncovalent, interaction exists between these two peptides in the rickettsial membrane; this is also a likely explanation for Anacker's observation (discussed above) that MAbs to the 120-kDa protein also allowed for the precipitation of a 32-kDa peptide. Protein sequences bracketing the cleavage site are preserved in *R. conorii, R. rhipicephali, R. montana, R. akari, R. australis, R. sibirica,* and the avirulent *R. rickettsii* Iowa strain. The Iowa strain is a derivative (natural mutant) of a once-virulent strain. Interestingly, posttranslational cleavage of its rOmpB was demonstrated to occur at reduced levels as compared to virulent *R. rickettsii.* Does an inability to process rOmpB lead to a decrease in virulence? The answer will hopefully be found after further investigation.

2.4. 17-kDa Rickettsia Genus-Common Protein

A gene from the Sheila Smith strain of *R. rickettsii,* encoding a 17-kDa protein, was cloned and expressed in *E. coli.*[36,37] Primer extension analysis revealed that the +1 mRNA start site was the same for both *E. coli* and *R. rickettsii.*[37] Interestingly, the recombinant protein was labeled in the presence of [³H]palmitate and [³H]glycerol, thus indicating lipid modification.[37] A surface location was established for this protein by the demonstration that antibodies, raised against the recombinant protein, reacted to the surface of *R. rickettsii* in immunoelectron microscopy.[38] Monospecific polyclonal antibodies to a portion of this protein reacted in Western blot analyses to 17-kDa proteins of *R. rickettsii, R. akari, R. conorii, R. belii,* as well as the typhus rickettsias *R. prowazekii, R. typhi,* and *R. canada.*;[39] these antibodies were not reactive with members of other genera including *Bacillus, Proteus, Neisseria, Escherichia,* and *Chlamydia.* The genes encoding this protein in *R. rickettsii, R. conorii, R. prowazekii, R. typhi,*[39] and *R. japonica*[40] have been sequenced and compared. The deduced amino acid sequence of the *R. conorii* protein was 100% homologous to that of *R. rickettsii*, whereas the *R. prowazekii* and *R. typhi* analogues were 91.8% homologous to the *R. rickettsii* protein; the *R. japonica* gene sequence was 98.5% homologous to the *R. rickettsii* gene. Given the high degree of this protein's conservation among *Rickettsia* spp., it is tempting to suggest that it has a function unique to this genus and closely related organisms.

3. SURFACE COMPONENTS OF A NONPROTEINACEOUS COMPOSITION

3.1. Lipopolysaccharide

Throughout the literature covering the antigens of SFG rickettsiae, there have been allusions to the presence of an LPS-like surface antigen. Most of these observations have come from investigations with MAbs.[9,13] These authors described a protease-resistant antigen that, on Western-blot analyses with some MAbs, presented in a "ladderlike" arrangement; Anacker *et al.* described this ladder arrangement as being similar to the pattern exhibited by the smooth LPS of enteric bacteria. None of the MAbs to the rickettsial LPS-like antigen have been shown to be protective in either the MPT or the guinea pig model of infection. This is in contrast to the fact that some MAbs to rOmpA and rOmpB do render passive protection. One must conclude that the ability to bind to any surface-exposed epitope is not enough to allow a MAb to be considered "protective." The question remains, what is this LPS-like antigen?

Only one study has extensively explored the chemical composition of SFG rickettsiae LPS.[41] LPS was extracted from the *Rickettsia* TT-118 SFG strain and the Japanese spotted fever group (*R. japonica*) Katayama strain. The major components of these LPSs were KDO, neutral sugars (an unknown sugar, ribose, and rhamnose), phosphate, glucosamine, quinovosamine, and palmitic acid. Unlike the LPSs found in enteric bacteria, no heptose or β-hydroxy fatty acids were detected in these LPSs. On Western blot and silver stain detection, the purified LPSs displayed the same ladderlike banding pattern described by Anacker *et al.*[9]

3.2. Microcapsular and Slime Layers

Microcapsular (MCL) and slime (SL) layers have been described for both the SFG and TG rickettsiae. Information on the SFG rickettsiae is scant. These layers have been visualized through electron microscopy and their chemical compositions have only been implied through differential staining with osmium tetroxide, ruthenium red, and methenamine silver, the latter two being specific for polysaccharides. Silverman *et al.*[42] described the MCL on *R. rickettsii* as consisting of small electron-dense projections extending from the cell wall (referred to in this chapter as the outer membrane). These authors described the SL as an "electron-lucent, halo-like zone," polysaccharide in composition, surrounding the organisms. This structure was reminiscent of capsules that have been identified on the surfaces of many bacteria. In consideration of electron microscopy and ruthenium red staining results, Hayes and Burgdorfer[43] reported the MCL and SL to be composed of acidic protein and polysaccharide. They also reported the presence of these layers on pathogenic rickettsiae but not on the nonvirulent 369-C isolate. The functions of these layers are not known. However, Hayes and Burdorfer did report that these layers, on *R. rickettsii* in ticks, become more discernible entities while the ticks feed. It is known that Rickettsiae in ticks undergo a transition from a relatively avirulent state to a virulent one during feeding. Therefore, Hayes and Burdorfer suggested that the changes in the MCL and SL may be a part of the virulence enhancement exhibited by these organisms.

4. CLOSING REMARKS

It is apparent from the studies summarized above that much is known about the general organization of the surfaces of SFG rickettsiae but little is known about the individual components. With further elucidation of the functions of these components, enhancement of our understanding of the pathogenesis of these organisms at the molecular level is certain. The scarcity of this knowledge makes the study of this unique group of organisms both interesting and challenging.

REFERENCES

1. Anderson, D. R., Hopps, H. E., Barile, M. F., and Bernheim, B. C., 1965, Comparison of the ultrastructure of several rickettsiae, ornithosis virus, and *Mycoplasma* in tissue culture, *J. Bacteriol.* **90:**1387–1404.
2. Silverman, D. J., and Wisseman, C. L., Jr., 1978, Comparative ultrastructural study on the cell envelopes of *Rickettsia prowazekii*, *Rickettsia rickettsii,* and *Rickettsia tsutsugamushi*, *Infect. Immun.* **21:**1020–1023.
3. Tzianabos, T., Palmer, E. L., Obijeski, J. F., and Martin, M. L., 1974, Origin and structure of the group-specific, complement-fixing antigen of *Rickettsia rickettsii*, *Appl. Microbiol.* **28:**481–488.
4. Obijeski, J. F., Palmer, E. L., and Tzianabos, T., 1974, Proteins of purified rickettsiae, *Microbios* **11:**61–76.
5. Feng, H. M., Kirkman, C., and Walker, D. H., 1996, Radioimmunoprecipitation of [35S] methionine-radiolabeled proteins of *Rickettsia conorii* and *Rickettsia rickettsii*, *J. Infect. Dis.* **154:**717–721.
6. Osterman, J. V., and Eisemann, C. S., 1978, Surface proteins of typhus and spotted fever group rickettsiae, *Infect. Immun.* **21:**866–873.
7. Anacker, R. L., Philip, R. N., Williams, J. C., List, R. H., and Mann, R. E., 1984, Biochemical and immunochemical analysis of *Rickettsia rickettsii* strains of various degrees of virulence, *Infect. Immun.* **44:**559–564.
8. Williams, J. C., Walker, D. H., Peacock, M. G., and Stewart, S. T., 1986, Humoral immune response to Rocky Mountain spotted fever in experimentally infected guinea pigs: Immu-

noprecipitation of lactoperoxidase ^{125}I-labeled proteins and detection of soluble antigens of *Rickettsia rickettsii*, *Infect. Immun.* **52:**120–127.

9. Anacker, R. L., List, R. H., Mann, R. E., Hayes, S. F., and Thomas, L. A., 1985, Characterization of monoclonal antibodies protecting mice against *Rickettsia rickettsii*, *J. Infect. Dis.* **151:**1052–1060.

10. Lange, J. V., and Walker, D. H., 1984, Production and characterization of monoclonal antibodies to *Rickettsia rickettsii*, *Infect. Immun.* **46:**289–294.

11. Anacker, R. L., List, R. H., Mann, R. E., and Wiedbrauk, D. L., 1986, Antigenic heterogeneity in high- and low-virulence strains of *Rickettsia rickettsii* revealed by monoclonal antibodies, Infect. Immun. **51:**653–660.

12. Anacker, R. L., McDonald, G. A., List, R. H., and Mann, R. E., 1987, Neutralizing activity of monoclonal antibodies to heat-sensitive and heat-resistant epitopes of *Rickettsia rickettsii* surface proteins. *Infect. Immun.* **55:**825–827.

13. Feng, H. M., Walker, D. H., and Wang, J. G., 1987, Analysis of T-cell-dependent and -independent antigens of *Rickettsia conorii* with monoclonal antibodies, *Infect. Immun.* **55:**7–15.

14. Li, H., Lenz, B., and Walker, D. H., 1988, Protective monoclonal antibodies recognize heat-labile epitopes on surface proteins of spotted fever group rickettsiae, *Infect. Immun.* **56:**2587–2593.

15. Walker, D. H., Liu, Q., Yu, X., Li, H., Taylor, C., and Fent, H., 1992, Antigenic diversity of *Rickettsia conorii*, *Am. J. Trop. Med. Hyg.* **47:**78–86.

16. Uchida, T., Tashiro, F., Funato, T., and Kitamura, Y., 1986, Isolation of a spotted fever group rickettsia from a patient with febrile exanthematous illness in Shikoku, Japan, *Microbiol. Immunol.* **30:**1323–1326.

17. Uchida, T., Uchiyama, T., and Koyama, A. H., 1988, Isolation of spotted fever group rickettsiae from humans in Japan, *J. Infect. Dis.* **158:**664–665.

18. Uchida, T., Yu, X., Uchiyama, T., and Walker, D. H., 1989, Identification of a unique spotted fever grop rickettsia from humans in Japan, *J. Infect. Dis.* **159:**1122–1126.

19. Uchiyama, T., Uchida, T., and Walker, D. H., 1990, Species-specific monoclonal antibodies to *Rickettsia japonica*, a newly identified spotted fever group rickettsia, *J. Clin. Microbiol.* **28:**1177–1180.

20. Uchiyama, T., Uchida, T., and Walker, D. H., 1994, Analysis of major surface polypeptides of *Rickettsia japonica*, *Microbiol. Immunol.* **38:**575–579.

21. Gilmore, R. D., Jr., Cieplak, W., Jr., Policastro, P. F., and Hackstadt, T., 1991, The 120 kilodalton outer membrane protein (rOmp B) of *Rickettsia rickettsii* is encoded by an unusually long open reading frame: Evidence for protein processing from a large precursor, *Mol. Microbiol.* **5:**2361–2370.

22. McDonald, G. A., Anacker, R. L., and Garjian, K., 1987, Cloned gene of *Rickettsia rickettsii* surface antigen: Candidate vaccine for Rocky Mountain spotted fever, *Science* **235:**83–85.

23. McDonald, G. A., Anacker, R. L., Mann, R. E., and Milch, L. J., 1988, Protection of guinea pigs from experimental Rocky Mountain spotted fever with a cloned antigen of *Rickettsia rickettsii*, *J. Infect. Dis.* **158:**228–231.

24. Anacker, R. L., Philip, R. N., Casper, E., Todd, W. J., Mann, R. E., Johnston, M. R., and Nauck, C. J., 1983, Biological properties of rabbit antibodies to a surface antigen of *Rickettsia rickettsii*, *Infect. Immun.* **40:**292–298.

25. Anderson, B. E., McDonald, G. A., Jones, D. C., and Regnery, R. L., 1990, A protective protein antigen of *Rickettsia rickettsii* has tandemly repeated, near-identical sequences, *Infect. Immun.* **58:**2760–2769.

26. Crocquet-Valdes, P. A., Weiss, K., and Walker, D. H., 1994, Sequence analysis of the 190-kDa antigen-encoding gene of *Rickettsia conorii* (Malish 7 strain), *Gene* **140:**115–119.

27. Gilmore, R. D., 1993, Comparison of the rompA gene repeat regions of rickettsia reveals species-specific arrangements of individual repeating units, *Gene* **125:**97–102.

28. Walker, D. H., Feng, H. M., Saada, J. I., Crocquet-Valdes, P., Radulovic, S., Popov, V. L., and Manor, E., 1995, Comparative antigenic analysis of spotted fever group rickettsiae from Israel and other closely related organisms, *J. Trop. Med. Hyg.* **52:**569–576.
29. Li, H., Crocquet-Valdes, P. A., and Walker, D. H., 1994, rOmpA is a critical protein for the adhesion of *Rickettsia rickettsii* to the host cell, *ASM*, Las Vegas, p. 100.
30. Gilmore, R. D., Joste, N., and McDonald, G. A., 1989, Cloning, expression and sequence analysis of the gene encoding the 120 kDa surface-exposed protein of *Rickettsia rickettsii, Mol. Microbiol.* **3:**1579–1586.
31. Policastro, P. F., and Hackstadt, T., 1994, Differential activity of *Rickettsia rickettsii ompA* and *ompB* promoter regions in a heterologous reporter gene system, *Microbiology* **140:**2941–2949.
32. Dasch, G. A., Samms, J. R., and Williams, J. C., 1981, Partial purification and characterization of the major species-specific protein antigens of *Rickettsia typhi* and *Rickettsia prowazekii* identified by rocket immunoelectrophoresis *Infect. Immun.* **31:**276–288.
33. Carl, M., Dobson, M. E., Ching, W. M., and Dasch, G. A., 1990, Characterization of the gene encoding the protective paracrystalline-surface-layer protein of *Rickettsia prowazekii:* Presence of a truncated identical homolog in *Rickettsia typhi, Proc. Natl. Acad. Sci. USA* **87:**8237–8241.
34. Ching, W.-M., Dasch, G. A., Carl, M., and Dobson, M. E., 1990, Structural analyses of the 120-kDa serotype protein antigens of typhus group rickettsiae, *Ann. N.Y. Acad. Sci.* **590:**334–351.
35. Hackstadt, T., Messer, R., Cieplak, W., and Peacock, M. G., 1992, Evidence for proteolytic cleavage of the 120-kilodalton outer membrane protein of rickettsiae: Identification of an avirulent mutant deficient in processing, *Infect. Immun.* **60:**159–165.
36. Anderson, B. E., Regnery, R. L., Carlone, G. M., Tzianabos, T., McDade, J. E., Fu, Z. Y., and Bellini, W. J., 1987, Sequence analysis of the 17-kilodalton-antigen gene from *Rickettsia rickettsii, J. Bacteriol.* **169:**2385–2390.
37. Anderson, B. E., Baumstark, B. R., and Bellini, W. J., 1988, Expression of the gene encoding the 17-kilodalton antigen from *Rickettsia rickettsii:* Transcription and posttranscriptional modification, *J. Bacteriol.* **170:**4493–4500.
38. Anderson, B. E., 1990, The 17-kilodalton protein antigens of spotted fever and typhus group rickettsiae, *Ann. N.Y. Acad. Sci.* **590:**326–333.
39. Anderson, B. E., and Tzianabos, T., 1989, Comparative sequence analysis of a genus-common rickettsial antigen gene. *J. Bacteriol.* **171:**5199–5201.
40. Furuya, T., Katayama, T., Yoshida, Y., and Kaiho, I., 1995, Specific amplification of *Rickettsia japonica* DNA from clinical specimens by PCR, *J. Clin. Microbiol.* **33:**487–489.
41. Amano, K., Fujita, M., and Suto, T., 1993, Chemical properties of lipopolysaccharides from spotted fever group rickettsiae and their common antigenicity with lipopolysaccharides from *Proteus* species, *Infect. Immun.* **61:**4350–4355.
42. Silverman, D. J., Wisseman, C. L., Jr., Waddell, A. D., and Jones, M., 1978, External layers of *Rickettsia prowazekii* and *Rickettsia rickettsii:* Occurence of a slime layer, *Infect. Immun.* **22:**233–246.
43. Hayes, S. F., and Burgdorfer, W., 1982, Reactivation of *Rickettsia rickettsii* in *Dermacentor andersoni* ticks: An ultrastructural analysis, *Infect. Immun.* **37:**779–785.
44. Bell, E. J., and Pickens, E. G., 1953, A toxic substance associated with the rickettsias of the spotted fever group, *J. Immunol.* **70:**461–472.
45. Moe, J. B., Mosher, D. F., Kenyon, R. H., White, J. D., Stookey, J. L., Bagley, L. R., and Fine, D. P., 1976, Functional and morphologic changes during experimental Rocky Mountain spotted fever in guinea pigs, *Lab. Invest.* **35:**235–245.
46. Li, H., Lenz, B., and Walker, D. H., 1988, Protective monoclonal antibodies recognize heat-labile epitopes on surface proteins of spotted fever group rickettsiae, *Infect. Immun.* **56:**2387–2593.

6

Oxidative Cell Injury and Spotted Fever Group Rickettsiae

DAVID J. SILVERMAN

1. INTRODUCTION

It has been generally accepted that the endothelial cell is the target cell in rickettsial infections such as Rocky Mountain spotted fever, epidemic typhus, murine typhus, and scrub typhus. These diseases are caused by obligate intracellular bacteria which belong either to the genus *Rickettsia* or to the genus *Orientia*. Although rickettsiae are capable of infecting and replicating within a variety of cultured eukaryotic cells, it was not until the successful cultivation of human endothelial cells *in vitro* in the early 1970s[1,2] that a unique opportunity arose to study these obligate intracellular bacteria in cells of the same species and ontogenic type as those found in human infection and disease. Although interpretative limitations exist in the extrapolation of *in vitro* observations to the actual *in vivo* condition, the study of rickettsia-infected endothelial cells is still likely to yield important information regarding clinical manifestations of human disease. For this reason, current studies in our laboratory have been confined almost exclusively to the human endothelial cell. The current focus of our laboratory is directed toward mechanisms of rickettsia-induced endothelial cell injury. The experimental data presented in the following sections suggest that injury caused

DAVID J. SILVERMAN • Department of Microbiology and Immunology, University of Maryland School of Medicine, Baltimore, Maryland 21201.

Rickettsial Infection and Immunity, edited by Anderson *et al*. Plenum Press, New York, 1997.

by *R. rickettsii* may be initiated by reactive oxygen species produced either during internalization of the bacteria or during intracellular growth and replication.

2. INFECTION OF HUMAN ENDOTHELIAL CELLS BY *R. RICKETTSII:* LIGHT MICROSCOPIC OBSERVATIONS

R. rickettsii is capable of infecting many different types of tissue culture cells *in vitro*. Perhaps the best and most comprehensive light microscopic characterization of an infection by this species of rickettsia has been accomplished by Wisseman *et al.*[3] in secondary chicken embryo fibroblasts and in mouse L-929 fibroblasts. Results of their study have shed light on some interesting and unique properties of *R. rickettsii* including the observation that these bacteria apparently are capable of traversing the host membrane from the cytosolic side as well as from the cell exterior. This mobility trait could explain the rapid cell-to-cell spread of *R. rickettsii* compared with *R. prowazekii*.[3-5] and, if corroborated in endothelial cells, may be related to the relatively short incubation period and rapid progression of Rocky Mountain spotted fever in humans.

Human endothelial cells in culture are readily infected by *R. rickettsii*,[6] and the progression of the infection is, for the most part, similar to that observed in other cell types. The percentage of cells infected and the number of rickettsiae per infected cell are dose and time dependent. The main differences in the infectivity of endothelial cells versus other cell types such as Vero (monkey kidney epithelial) are the lower average number of rickettsiae per endothelial cell at specific time points after infection, and a slightly slower rate of cell-to-cell spread. These differences may be attributed either to some undefined biochemical or metabolic inadequacy on the part of the endothelial cell or, possibly, to some unique form of acute cell injury caused by *R. rickettsii* which renders the endothelial cell less capable of supporting maximal rickettsial growth.

As has been shown previously in cultured cells other than endothelial cells,[3,7,8] *R. rickettsii*, once internalized, replicates in both the nucleus and cytoplasm of the infected cell. Cytoplasmic and intranuclear replication of this rickettsia appear to occur in endothelial cells as well,[6] although nuclear localization in all cell types studied so far seems to be limited to a small percentage of the total cell population ($<5\%$), and the significance of the organisms' location in the nucleus, if any, is unknown. Figure 1 shows human endothelial cells from a representative population 96–120 hr postinfection with *R. rickettsii*. As the photomicrographs indicate, these bacteria display a variety of interesting growth patterns within endothelial cells.

These light microscopic studies have demonstrated the capacity for infection of cultured human endothelial cells with *R. rickettsii*, and thus, the potential for

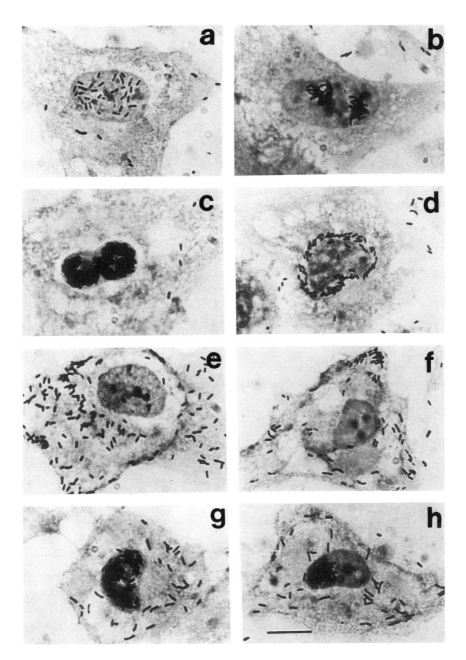

FIGURE 1. *R. rickettsii* in human umbilical vein endothelial cells 96–120 hr postinfection showing the various distribution patterns of rickettsiae within infected cells: predominantly nuclear involvement (a–c), cytoplasmic localization (d–f), and both nuclear and cytoplasmic distribution (g and h). Note masses of rickettsiae in some nuclei and diffuse distribution in cytoplasm. Bar = 10 μm. (From Silverman, D. J., and Bond, S. B., 1984, Infection of human vascular endothelial cells by *Rickettsia rickettsii, J. Infect. Dis.* **149:**201, copyright, The University of Chicago Press.)

these cells to be used as an *in vitro* model for the study of rickettsia–host cell interactions.

3. ELECTRON MICROSCOPY OF HUMAN ENDOTHELIAL CELLS INFECTED BY *R. RICKETTSII:* MORPHOLOGICAL MANIFESTATIONS OF CELL INJURY

As part of a continuing effort to understand the basis for the relationship that exists between bacteria belonging to the genus *Rickettsia* and eukaryotic cells, our laboratory previously undertook a comparative electron microscopic study of chicken embryo cells infected by *R. rickettsii* and *R. prowazekii*.[9–11] These studies, which were carried out primarily to evaluate intracellular changes following infection, showed significant differences between the two organisms. *R. prowazekii,* about which Wisseman and Waddell[4] previously had reported, grew to large numbers within infected cells, and caused only modest morphological changes to cytoplasmic organelles and membranes.[10] Cells infected by *R. rickettsii,* on the other hand, showed signs of injury as early as 48 hr postinfection with more prominent changes occurring after an additional 48 to 72 hr. These changes, which were manifested primarily as dilatation of the rough-surfaced endoplasmic reticulum outer nuclear envelope complex (RER-ONE), presumably are related to increased water and sodium ion influx as a result of enhanced membrane permeability.

Following characterization of the infection and replication of *R. rickettsii* in human endothelial cells by light microscopy, infected cells were subsequently examined by electron microscopy to determine whether the membrane changes observed in other cell types infected by this rickettsia also occurred in endothelial cells.[12] Examination of infected cell populations confirmed the same series of changes to the RER-ONE. Figure 2 is an electron micrograph representative of an endothelial cell 24 hr after infection with *R. rickettsii.* The rickettsiae typically are found free in the cytosol, and there is no morphological evidence of injury to the host. As the infection progresses, however, dilatation of the RER-ONE becomes increasingly more prominent (Figs. 3–5), with presumed progression to eventual cell lysis. What is truly remarkable is the relatively few rickettsiae (as monitored in parallel light microscopic samples) capable of causing these dramatic changes to the host. This characteristic of the infection led us to hypothesize that injury to endothelial cells might occur by some cytotoxic phenomenon, possibly biochemical anomalies occurring as a result of altered host cell metabolism or even as a direct result of rickettsial metabolism.

Following this morphological study, potential mechanisms of cell injury that would be consistent with the electron microscopic observations were considered. The first of these was the possible involvement of the enzyme phospholipase A,

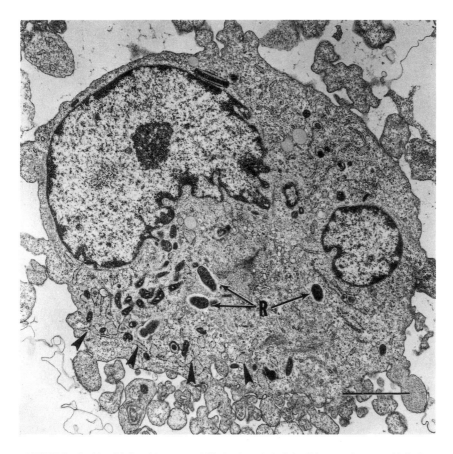

FIGURE 2. *R. rickettsii*-infected human umbilical vein endothelial cell in an early stage of infection, showing organisms (R) in the cytosol. Rickettsiae are easily distinguished from mitochondria because of the "halo" that surrounds them and that appears to prevent contact between the organisms and the host cytosol. Arrowheads identify RER. Bar = 2 μm. (From Silverman, D. J., 1984, *Rickettsia rickettsii*-induced cellular injury of human vascular endothelium in vitro, *Infect. Immun.* **44:**545, copyright American Society for Microbiology.)

which Winkler and Miller[13] previously had proposed is involved in both the internalization of and escape of *R. prowazekii* from infected host cells. Their hypothesis was based on the detection of glycerophospholipids released from host cells following exposure to sizable rickettsial inocula. Although similar experiments were not carried out with *R. rickettsii*, the changes in the internal membranes of infected cells alone, warranted consideration of membrane-modifying enzymes such as phospholipase as potential candidates for cell injury. A consider-

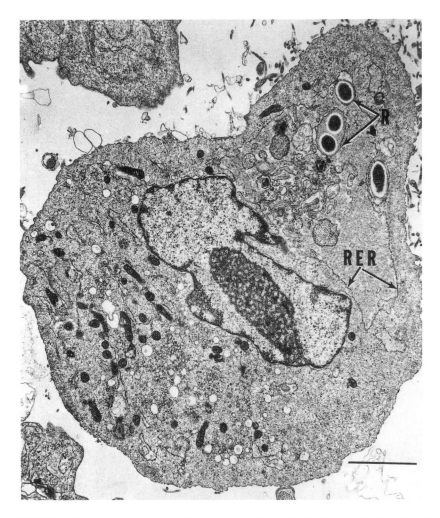

FIGURE 3. Beginning states of cytopathic change induced in an *R. rickettsii*-infected endothelial cell. Rough-surfaced endoplasmic reticulum (RER) shows substantial dilatation, and rickettsiae (R), which typically are not host membrane associated, are circumscribed by the expanding internal membranes. Bar = 2 μm. (From Silverman, D. J., 1984, *Rickettsia rickettsii*-induced cellular injury of human vascular endothelium in vitro, *Infect. Immun.* **44:**545, copyright American Society for Microbiology.)

able reservation in advancing this idea was that *R. prowazekii* does not cause the same changes to the RER-ONE as does *R. rickettsii.*

One type of cell injury that is consistent with these types of changes to intracellular membranes, however, occurs in cells suffering from oxidative stress.[14–16] As a result, we turned our attention to reactive oxygen species and their potential role in cell injury caused by *R. rickettsii.*

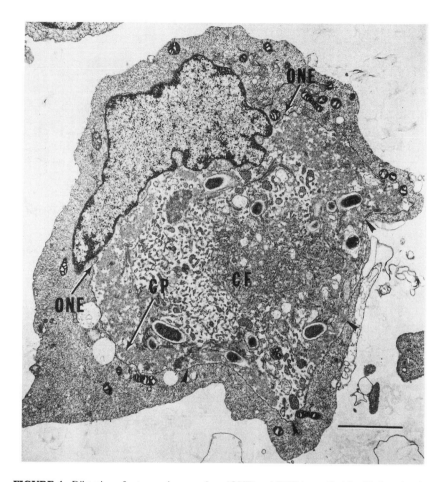

FIGURE 4. Dilatation of outer nuclear envelope (ONE) and RER in an *R. rickettsii*-infected endothelial cell, creating a large intracellular cisterna (outlined by arrowheads). Host cytoplasmic fragments (CF), intracisternal protein (CP), and membrane-associated rickettsiae, as well as an occasional mitochondrion, can be seen within the cisterna. Bar = 2μm. (From Silverman, D. J., 1984, *Rickettsia rickettsii*-induced cellular injury of human vascular endothelium in vitro, *Infect. Immun.* **44:**545, copyright American Society of Microbiology.)

4. BIOCHEMICAL EVIDENCE IN SUPPORT OF A HYPOTHESIS OF OXIDATIVE CELL INJURY CAUSED BY RICKETTSIAL INFECTION: INTRACELLULAR PEROXIDES

In light of the above electron microscopic observations, a timely correlation was sought between morphological manifestations of cell injury and the presence of intracellular oxidants that could initiate these changes. Several reactive oxygen

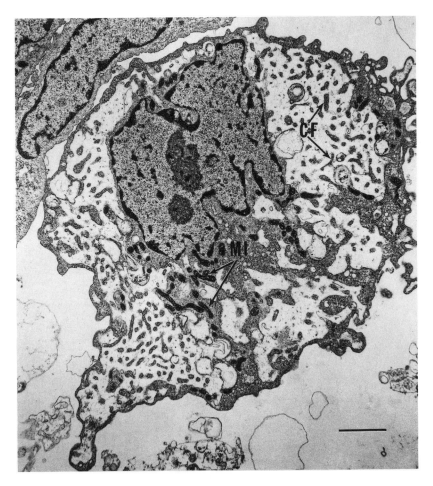

FIGURE 5. Advanced stage of RER dilatation in an *R. rickettsi*-infected endothelial cell. Note the reduced electron opacity in the cisterna, resulting from the loss of fine granular intracisternal protein. Physically intact mitochondria (MI) appear normal. CF, cytoplasmic fragments. Bar = 2 μm. (From Silverman, D. J., 1984, *Rickettsia rickettsii*-induced cellular injury of human vascular endothelium in vitro, *Infect. Immun.* **44**:545, copyright American Society for Microbiology.)

species, including superoxide and hydrogen peroxide, are produced during normal cellular metabolism. Typically, the levels of these oxidants are kept in check by host scavenger systems which prevent cellular damage. However, if cellular antioxidant systems are compromised in some manner, cells may become exposed to unusually high levels of reactive oxygen species with subsequent cell injury and death.

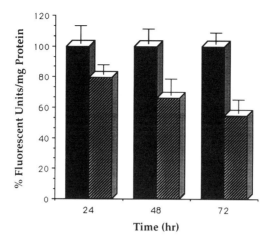

FIGURE 6. Intracellular peroxide levels in human endothelial cells infected by *R. rickettsii* measured at 24, 48, and 72 hr postinfection. Solid bar, infected cell population; hatched bar, uninfected control cell population. Data are means ± SE of at least 15 determinations at each time point. (From Silverman, D. J., and Santucci, L. A., 1988, Potential for free radical-induced lipid peroxidation as a cause of endothelial cell injury in Rocky Mountain spotted fever, *Infect. Immun.* **56:**3110, copyright American Society for Microbiology.)

As a first step in determining the status of antioxidant defenses in endothelial cells infected by *R. rickettsii*, the levels of intracellular peroxides were monitored using the probe 5- (and 6-) carboxy-2',7'-dichlorofluorescein diacetate.[17] This probe is deacetylated on passage through the plasma membrane and is oxidized to the fluorescent compound 5- (and 6-) carboxy-2',7'-dichlorofluorescein by oxidants such as hydrogen peroxide and lipid peroxides.[18] Experimental results of our study showed that significantly higher levels of peroxides were produced in populations of infected cells than in uninfected cells[17] (Fig. 6).

Another oxidant, the hydroxyl radical, generated by the interaction of superoxide and hydrogen peroxide,[19] is one of the more biologically active oxygen radicals, and is a potent initiator of lipid peroxidation through the formation of lipid hydroperoxides. To determine whether this radical might contribute to increased peroxide levels in infected cells, we used the iron chelator desferrioxamine which blocks the formation of the hydroxyl radical.[19] Desferrioxamine effectively reduced the levels of intracellular peroxides in infected cells,[17] suggesting that lipid peroxides may, indeed, accumulate in endothelial cells infected by *R. rickettsii*.

5. INDUCTION OF SUPEROXIDE DISMUTASE AND RELEASE OF SUPEROXIDE DURING INFECTION OF ENDOTHELIAL CELLS BY *R. RICKETTSII*

Hydrogen peroxide is formed by the dismutation of superoxide according to the following reaction:

$$2O_2^- + 2H^+ \xrightarrow{\text{SOD}} O_2 + H_2O_2$$

The enzyme superoxide dismutase, which catalyzes this reaction, is present in cells and affords them protection from the toxic effects of the superoxide radical generated during normal cellular metabolism. Cells respond to elevated levels of intracellular superoxide by increased synthesis of superoxide dismutase to neutralize this radical and to prevent oxidative injury.[20-22]

As a follow-up to the studies on intracellular peroxides, the levels of superoxide dismutase in endothelial cells were examined following infection by *R. rickettsii*.[23] Levels of this enzyme increased sharply within the first 2 to 6 hr after exposure of the cells to the rickettsiae, and remained elevated for at least 48 hr (Fig. 7). Pretreatment of cells with cycloheximide prevented increases in superoxide dismutase, indicating that the enzyme was synthesized *de novo* following rickettsial infection.

Because elevated levels of superoxide provide the typical stimulus for increased synthesis of superoxide dismutase, experiments were carried out to localize and identify the source of this oxygen radical. Conventional methods for detection of intracellular superoxide are difficult to use. However, sensitive spectrophotometric and electron paramagnetic resonance (EPR) techniques exist that can be used for the detection and measurement of extracellular superoxide. Examination of the culture supernatants of endothelial cells shortly after exposure to *R. rickettsii* showed that release of superoxide was dependent, to the extent tested, on the concentration of the rickettsial inoculum (Fig. 8). Superoxide was detected 1 hr after exposure to a rickettsial inoculum containing six rickettsiae per endothelial cell, but not with an inoculum containing one rickettsia per cell. At 24 hr postinfection, increases in superoxide were not detected with either size inoculum, suggesting that during cell-to-cell spread there may not be a sufficiently high enough number of coincident entry events to detect release of superoxide.

Pretreatment of cells with cytochalasin D, an inhibitor of rickettsial entry, reduced the amount of measurable superoxide, suggesting that release of this radical is closely related to the internalization of *R. rickettsii*. The eukaryotic protein synthesis inhibitor cycloheximide, on the other hand, when added to cells prior to exposure to *R. rickettsii*, resulted in an increase in superoxide release (Fig. 9).

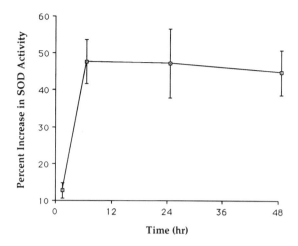

FIGURE 7. Percent increase in SOD activity in human endothelial cells following infection with *R. rickettsii.* Enzyme activity was measured at 0, 6, 24, and 48 hr postinfection and is expressed as the percent increase in SOD activity in infected cells compared with that in uninfected controls. Data are means ± SE of at least four separate experiments at each time point. (From Santucci, L. A., Gutierrez, P. L., and Silverman, D. J., 1992, *Rickettsia rickettsii* induces superoxide radical and superoxide dismutase in human endothelial cells, *Infect. Immun.* **60:**5113, copyright American Society for Microbiology.)

That injury may be initiated soon after rickettsial exposure is supported by recent studies in our laboratory using fluorescence-activated cell sorter (FACS) analysis (in preparation) that demonstrate the presence of intracellular oxidant activity in a population of endothelial cells as early as 4 hr postinfection, much earlier than previously thought.[17] These recent results are in line with our previous finding that showed early induction of superoxide dismutase following infection by *R. rickettsii.*[23]

How does extracellularly generated superoxide affect *de novo* synthesis of intracellular superoxide dismutase? Presumably, to stimulate additional synthesis of this enzyme, superoxide must enter the cell. This anion could enter cells by one of two mechanisms: (1) it could enter together with the rickettsiae during internalization of the organism or (2) it could enter cells through anion channels in the plasma membrane.[24] Either mechanism, theoretically, could trigger increased *de novo* synthesis of superoxide dismutase. An alternative, but unvalidated possibility, is that superoxide also might be generated intracellularly following infection, which would lead directly to induction of superoxide dismutase synthesis.

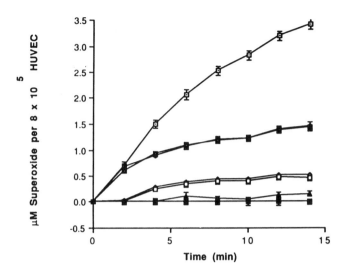

FIGURE 8. Superoxide radical measurement in supernatant fluids of human endothelial cells infected with *R. rickettsii*. Superoxide was measured by monitoring ferricytochrome c reduction spectrophotometrically at 550 nm. In the infected populations, endothelial cells were exposed to an average of either one rickettsia per cell or six rickettsiae per cell for 1 hr. SOD was added to parallel supernatants to ensure the specificity of the reaction. Data are means ± SE of eight determinations at each time point. ⊡, six *R. rickettsii* per cell; ◆ one *R. rickettsii* per cell; ▪, no *R. rickettsii*; ◆, purified *R. rickettsii*; ▦, six *R. rickettsii* per cell plus SOD; ▯, one *R. rickettsii* per cell plus SOD; ▲, SOD without rickettsiae. (From Santucci, L. A., Gutierrez, P. L., and Silverman, D. J., 1992, *Rickettsia rickettsii* induces superoxide radical and superoxide dismutase in human endothelial cells, *Infect. Immuno.* **60:**5113, copyright American Society of Microbiology.)

6. PROTECTION OF *R. RICKETTSII*-INFECTED ENDOTHELIAL CELLS BY HEPARIN

The glycosaminoglycan heparin is routinely used in the culture of human endothelial cells because it reduces cell doubling time and increases cell population size.[25] Heparin recently has been shown to protect arterial endothelial cells from artificially induced superoxide radicals.[26] In light of this observation, the effects of heparin on endothelial cell injury in our system were examined. Preliminary experiments were carried out to ensure that heparin was not toxic to *R. rickettsii*. At the concentration of heparin typically used for culturing endothelial cells (50 μg/ml), neither rickettsial uptake nor the kinetics of intracellular rickettsial growth was compromised.[27]

Having established a commonality for the growth of *R. rickettsii*, in the presence and absence of heparin, an assessment could be made as to the efficacy of heparin to reduce or to abrogate endothelial cell injury. Several parameters

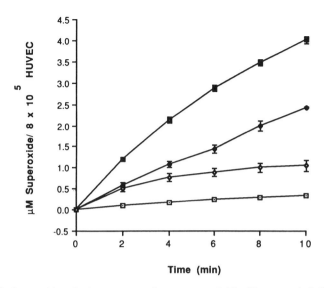

FIGURE 9. Superoxide radical measurement in supernatant fluids of human endothelial cells that were pretreated with either cycloheximide or cytochalasin D for 30 min prior to infection with *R. rickettsii*. In these experiments, endothelial cells were exposed to an average of six rickettsiae per cell for 1 hr. Data are means ± SE of at least 10 determinations at each time point. □, no rickettsiae; ◆, *R. rickettsii;* ■, 1 µg/ml cycloheximide plus *R. rickettsii;*◆, 0.5 µg/ml cytochalasin D plus *R. rickettsii*. (From Santucci, L. A., Gutierrez, P. L., and Silverman, D. J., 1992, *Rickettsiae rickettsii induces superoxide radical and superoxide dismutase in human endothelial cells, Infect. Immun.* **60:**5113, copyright American Society for Microbiology.)

were examined: (1) effect of heparin on the viability of infected endothelial cells; (2) effect of heparin on rickettsial plaque formation (ability to cause cytopathic effect); (3) effect of heparin on dilatation of intracellular membranes; and (4) effect of heparin on intracellular peroxide levels. Viability of cells was tested daily for 3 days using trypan blue dye exclusion with lactic dehydrogenase (LDH) release employed to assess cell injury. Heparin clearly offered protection to infected cells as early as 24 hr after exposure to *R. rickettsii* as indicated by reduced trypan blue staining and reduction in LDH release. When heparin was incorporated into plaque assay experiments, it caused a significant reduction in the size of plaques (>40%). Electron microscopic examination of infected cell cultures at 48 and 72 hr postinfection also clearly showed that heparin was effective in diminishing dilatation of the RER-ONE membrane complex. Lastly, the presence of heparin significantly reduced the levels of intracellular peroxides in infected cell cultures.

How could heparin exert its effect in this system? One possibility that was not fully appreciated at the time of our initial observations is the ability of heparin to remove extracellular superoxide dismutase from the surface of endothelial cells.[28,29] Extracellular superoxide dismutase apparently exists in three forms, [A], [B], and [C]. [A] has no affinity for heparin and, together with [B], which has a weak affinity, is thought to play a role in converting superoxide to hydrogen peroxide in the plasma of blood vessels. [C], on the other hand, which has a high affinity for the glycosaminoglycans (heparan sulfates) of the glycocalyxes of endothelial cells, has a strong affinity for heparin. When heparin is added to cell cultures, it competes with cell-associated superoxide dismutase and removes the latter from the cell surface.[29] The consequence is that the capacity for conversion of superoxide to the more toxic and stable hydrogen peroxide at the cell surface is reduced, and thus, there is less potential for the latter to freely enter cells. If heparin is internalized together with the rickettsiae, it may also exert some effect intracellularly, although there are no current experimental data to support this.

7. GLUTATHIONE: DEPLETION DURING RICKETTSIAL INFECTION OF ENDOTHELIAL CELLS AND ROLE IN PROTECTION AGAINST OXIDATIVE INJURY

Thiols, and in particular, glutathione, an integral part of the glutathione–redox cycle, play an important role in detoxifying intracellular peroxides and in averting oxidative injury. Reduced glutathione (GSH), which comprises a major portion of intracellular thiols, is oxidized in the presence of peroxides, and is regenerated through the action of glutathione reductase and NADPH. Generation of peroxides in excess of that which can be neutralized by endogenous levels of glutathione can result in cell injury. As it was previously shown that abnormally high levels of peroxides accumulate in endothelial cells following infection by *R. rickettsii*,[17] it was of interest to determine how these elevated peroxides affected intracellular levels of reduced glutathione (thiols).

In routine experiments, thiol levels in infected endothelial cells were reduced by about 50% within the first 24 hr postinfection and declined further by 48 hr[30] (Fig. 10), suggesting a severe compromise to an important line of intracellular defense against oxidative injury. γ-Glutamylcysteine, the dipeptide precursor of glutathione, was used to restore thiol levels in infected cells. This dipeptide, unlike the intact glutathione molecule, is transported through the plasma membrane and can be utilized directly for glutathione synthesis.[31] When glutathione was restored in this fashion to infected cells, there was a notable effect on plaque formation. The size of plaques was markedly reduced in a dose-dependent manner when single additions of γ-glutamylcysteine were added to agarose overlays. These results, although preliminary, suggest a probable causal relationship between reduced glutathione levels and cell injury.

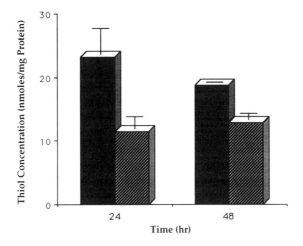

FIGURE 10. Comparative thiol concentrations in human umbilical vein endothelial cells 24 and 48 hr after infection by *R. rickettsii*. Data are averages ± SEM. Solid bars, uninfected endothelial cells; hatched bars, infected cells. (From Silverman, D. J., and Santucci, L. A., 1990, A potential role for thiols against cell injury caused by *Rickettsia rickettsii*, in: *Rickettsiology: Current Issues and Perspectives, Ann. N.Y. Acad. Sci.* **590:**111, copyright New York Academy of Sciences.)

8. ACTIVITY OF KEY ANTIOXIDANT ENZYMES INVOLVED IN PROTECTION AGAINST OXIDATIVE INJURY IN ENDOTHELIAL CELLS INFECTED BY *R. RICKETTSII*

In addition to thiol compounds such as glutathione, cells are dependent on antioxidant enzymes for protection against oxidative injury. Two of these enzymes, glutathione peroxidase and glucose-6-phosphate dehydrogenase (G-6-PD), are intimately linked with the glutathione–redox cycle and are important in the regeneration of reduced glutathione for detoxification of hydrogen peroxide and lipid hydroperoxides. A third enzyme, catalase, which acts to detoxify hydrogen peroxide, also plays a critical role in the protection of cells against oxidative stress. In a comparative study, the activity of all three enzymes was reduced in endothelial cells infected by *R. rickettsii*.[32] (Table I). The activity of catalase, which did not appreciably change during the first 24 hr postinfection, dropped by more than 50% of control values after 48 hr. The activity of glutathione peroxidase was markedly reduced shortly after infection, decreasing by about 40% after 24 hr, and by 65% after 48 hr. Lastly, G-6-PD, although showing only a modest decrease in activity by 24 hr postinfection, showed a striking drop (80%) after 48 hr.

TABLE I
Levels of Catalase, Glutathione Peroxidase, and G-6-PD in Human Endothelial Cells

	Level (mU/mg protein)[a] of					
	Catalase in		Glutathione peroxidase in		G-6-PD in	
Time (hr)	Uninfected cells	Infected cells	Uninfected cells	Infected cells	Uninfected cells	Infected cells
24	2139±337	1820±282	102.34±7.5	58.40±4.8	22.34±0.83	16.84±0.92
48	2720±250	1180±160	113.37±7.8	38.87±4.2	25.22±0.77	4.30±0.46

[a]Each value represents the mean ± SEM of three to five individual experiments. Infected samples had significantly lower values than did their respective controls, at a p value of <0.0001, except in the case of catalase at 24 hr, for which the difference is not significant.

The causes of the reduction in activity of these key enzymes in endothelial cells infected by *R. rickettsii* are not known. However, some precedents occur in the literature that allude to the susceptibility of catalase and glutathione peroxidase to oxidative inhibition by the superoxide radical.[33,34] G-6-PD also is susceptible to oxidative modification through carbonyl formation which inhibits enzymatic activity.[35] The decreased activity of G-6-PD likely has a direct correlation with the previously reported reduction in thiols[30] as this enzyme is coupled with the regeneration of reduced glutathione.

9. PHOSPHOLIPASE A: EVIDENCE FOR ITS PRESENCE ON *R. RICKETTSII*, AND PROSPECTS FOR ITS ROLE IN CELL INJURY

The enzyme phospholipase A appears to play an important role in rickettsial infection. Winkler and his colleagues[13,36,37] have proposed that this enzyme is involved both in the internalization and in the release of *R. prowazekii* from eukaryotic cells. Furthermore, Walker *et al.*[38] have shown that an inhibitor of phospholipase A reduces plaque formation by *R. rickettsii*, and attribute this reduction to inhibition of internalization of the organism. Recent studies from our laboratory provide additional evidence in support of phospholipase A's role in the internalization of *R. rickettsii*. Through the use of specific inhibitors of this enzyme and the immunodetection of epitopes on the rickettsial outer membrane that are cross-reactive with eukaryotic phospholipase A_2, there is now good evidence to suggest that the enzyme involved in the internalization of R. rickettsii is of rickettsial origin.[39,40]

What role could this enzyme play in cell injury caused by rickettsiae? The generation of superoxide at the surface of R. rickettsii-infected endothelial cells may come from the activation of an NADPH oxidase or similar enzyme as occurs in neutrophils. Although not normally considered professional phagocytes, endothelial cells can bind and ingest bacteria and can discriminate between bacterial strains.[41,42] Furthermore, it has been shown that a mutant of Salmonella, when exposed to endothelial cells, produces a greater release of superoxide (35 nmol per $3 - 10^6$ cells) than cells stimulated with phorbol myristate acetate or the calcium ionophore A23187.[41] Interestingly, the superoxide released from endothelial cells infected by R. rickettsii is 50 to 100 times higher than the above value, and from one-third the number of endothelial cells.[23]

NADPH oxidase is relatively inactive until cells are stimulated by some membrane-associated event. Once activated, the enzyme appears to be situated in the plasma membrane with the NADPH-binding site oriented toward the cytoplasm, and the oxygen-binding and superoxide-releasing component situated toward the exterior or phagocytic vesicle.[43,44] NADPH oxidase can be activated by detergents such as lysophospholipids, by-products of phospholipase A_2 digestion. Also, arachidonic acid, a by-product of both phospholipase A_2 and phospholipase C digestion of phospholipids, has detergent action and has been shown to activate NADPH oxidase.[45-49]

From the studies in our laboratory[39,40] and those of Winkler et al.[13,36,37] and Walker et al.[38] it is now generally accepted that phospholipases play a probable role in the internalization of rickettsiae. If this is true, then activation of an NADPH oxidase (or similar enzyme) by a phospholipase could explain the release of superoxide at the cell surface of endothelial cells infected by R. rickettsii. As our preference is that R. rickettsii may enter host cells by an "abortive phagocytosis" (abortive in the sense that the organism may never be fully enclosed by the membranes of a phagocytic vacuole), it is possible that during internalization, superoxide, and possibly other reactive oxygen species, leak into the cytosolic space where they may initiate a series of pathological sequelae.

A recent report[50] has shown that although phospholipase A_2 and an oxidant such as hydrogen peroxide alone may not be very effective in causing cell injury, the combination of both the enzyme and the oxidant, acting in tandem, resulted in significant cell damage. Although we have not looked specifically for a synergistic effect between phospholipase and hydrogen peroxide in the R. rickettsii-endothelial cell system, the individual components that comprise this combination have tentatively been identified, and therefore, the potential for mediation of cell injury.

10. SUMMARY

Morphological and biochemical evidence cited herein appear to support the hypothesis that reactive oxygen species play a role in the injury of endothelial cells by *R. rickettsii*. Cell injury may be initiated by the release of superoxide during rickettsial internalization, and subsequent injury may be caused either by this radical or by other reactive oxygen species. Although cells apparently respond appropriately to the generation of superoxide by increasing *de novo* synthesis of superoxide dismutase, normal antioxidant defenses such as detoxification of peroxides seem to be compromised as a result of infection. Thiol (glutathione) levels decrease within the first 24 hr following infection, and the activities of key antioxidant enzymes such as catalase, glutathione peroxidase, and G-6-PD are reduced, in some cases by more than 80% by 48 hr postinfection. These breaches in the integrity of antioxidant defenses may be sufficient to render cells susceptible to reactive oxygen species. Presently, evidence for this type of rickettsia-induced cell injury exists only for *R. rickettsii*. Further studies are necessary to determine if other organisms belonging to the genus *Rickettsia* also are capable of causing oxidant-mediated injury.

ACKNOWLEDGMENT. This work was supported by Public Health Service grant AI 17416 from the National Institutes of Allergy and Infectious Diseases.

REFERENCES

1. Jaffe, E. A., Hoyer, L. A., and Nachman, R. L., 1973, Synthesis of antihemophilic factor antigen by cultured human endothelial cells, *J. Clin. Invest.* **52**:2757–2764.
2. Gimbrone, M. A., 1976, Culture of vascular endothelium, *Prog. Hemostasis Thromb.* **3**:2745–2756.
3. Wisseman, C. L., Jr., Edlinger, E. A., Waddell, A. D., and Jones, M. R., 1976, Infection cycle of *Rickettsia rickettsii* in chicken embryo and L-929 cells in culture, *Infect. Immun.* **14**:1052–1064.
4. Wisseman, C. L., Jr., and Waddell, A. D., 1975, In vitro studies of rickettsia–host cell interactions: Intracellular growth cycle of virulent and attenuated *Rickettsia prowazekii* in chicken embryo cells in slide chamber cultures, *Infect. Immun.* **11**:1391–1401.
5. Wisseman, C. L., Jr., Waddell, A. D., and Silverman, D. J., 1976, In vitro studies on rickettsia–host cell interactions: Lag phase in intracellular growth cycle as a function of stage of growth of infecting *Rickettsia prowazekii*, with preliminary observations on inhibition of rickettsial uptake by host cell fragments, *Infect. Immun.* **13**:1749–1760.
6. Silverman, D. J., and Bond, S. B., 1984, Infection of human vascular endothelial cells by *Rickettsia rickettsii*, *J. Infect. Dis.* **149**:201–206.
7. Burgdorfer, W. R., Anacker, R. L., Bird, R. G., and Bertram, D. S., 1968, Intranuclear growth of *Rickettsia rickettsii*, *J. Bacteriol.* **96**:1415–1418.
8. Pinkerton, H., and Hass, G. M., 1932, Spotted fever I. Intranuclear rickettsiae in spotted fever studied in tissue culture, *J. Exp. Med.* **56**:151–156.
9. Silverman, D. J., and Wisseman, C. L., 1979, In vitro studies of rickettsia–host cell interactions: Ultrastructural changes induced by *Rickettsia rickettsii* infection of chicken embryo fibroblasts, *Infect. Immun.* **26**:714–727.

10. Silverman, D. J., Wisseman, C. L., Jr., and Waddell, A., 1980, In vitro studies of rickettsia–host cell interactions: Ultrastructural study of *Rickettsia prowazekii*-infected chicken embryo fibroblasts, *Infect. Immun.* **29:**778–790.
11. Silverman, D. J., Wisseman, C. L., Jr., and Waddell, A., 1981, Envelopment and escape of *Rickettsia rickettsii* from host cell membranes, in: *Rickettsiae and Rickettsial Diseases* (W. Burgdorfer and R. L. Anacker, eds.), Academic Press, New York, pp. 241–253.
12. Silverman, D. J., 1984, *Rickettsia rickettsii*-induced cellular injury of human vascular endothelium in vitro, *Infect. Immun.* **44:**545–553.
13. Winkler, H. H., and Miller, E. T., 1982, Phospholipase A and the interaction of *Rickettsia prowazekii* and mouse fibroblasts (L-929 cells), *Infect. Immun.* **38:**109–113.
14. Freeman, B. A., and Crapo, J. D., 1982, Biology of disease: Free radicals and tissue injury, *Lab. Invest.* **47:**412–426.
15. Smuckler, E. A., 1976, Alterations produced in the endoplasmic reticulum by carbon tetrachloride, *Panminerva Med.* **18:**292–303.
16. Trump, B. F., Croaker, B., and Mergner, W. J., 1971, The role of energy metabolism, ions, and water shifts in the pathogenesis of cell injury, in: *Cell Injury* (D. G. Scarpelli and B. F. Trump, eds.), The Upjohn Company, Kalamazoo, MI, pp. 1–72.
17. Silverman, D. J., and Santucci, L. A., 1988, Potential for free radical-induced lipid peroxidation as a cause of endothelial cell injury in Rocky Mountain spotted fever, *Infect. Immun.* **56:**3110–3115.
18. Cathcart, R., Schwiers, E., and Ames, B. N., 1983, Detection of picomole levels of hydroperoxides using a fluorescent dichlorofluorescein assay, *Anal. Biochem.* **134:**111–116.
19. Halliwell, B., and Gutteridge, J. M. C., 1986, Oxygen free radicals and iron in relation to biology and medicine: Some problems and concepts, *Arch. Biochem. Biophys.* **246:**501–524.
20. Fridovich, I., 1988, The biology of oxygen radicals: General concepts, in: *Oxygen Radicals and Tissue Injury* (B. Halliwell, ed.), The Upjohn Company, Kalamazoo, MI, pp. 1–5.
21. Fridovich, I., 1972, Superoxide radical and superoxide dismutase, *Acc. Chem. Res.* **5:**321–326.
22. Hassan, H. M., 1985, Manipulation of SOD levels in prokaryotes, in: *Handbook of Methods of Oxygen Radical Research* (R. A. Greenwald, ed.), CRC Press, Boca Raton, pp. 353–358.
23. Santucci, L. A., Gutierrez, P. L., and Silverman, D. J., 1992, *Rickettsia rickettsii* induces superoxide radical and superoxide dismutase in human endothelial cells, *Infect. Immun.* **60:**5113–5118.
24. Kontos, H. A., Wei, E. P., Ellis, E. F., Jenkins, L. W., Povlishock, J. T., Rowe, G. T., and Hess, M. L., 1985, Appearance of superoxide anion radical in cerebral extracellular space during increased prostaglandin synthesis in cats, *Circ. Res.* **57:**142–151.
25. Thornton, S. C., Mueller, S. N., and Levine, E. L., 1983, Human endothelial cells: Use of heparin in cloning and long-term serial cultivation, *Science* **227:**221–223.
26. Hiebert, L. M., and Liu, J., 1990, Heparin protects cultured arterial endothelial cells from damage by toxic oxygen metabolites, *atherosclerosis* **83:**47–51.
27. Silverman, D. J., Santucci, L. A., and Sekeyova, Z., 1991, Heparin protects human endothelial cells infected by *Rickettsia rickettsii*, *Infect. Immun.* **59:**4505–4510.
28. Adachi, T., Ohta, H., Hayashi, K., Hirano, K., and Marklund, S. L., 1992, The site of nonenzymic glycation of human extracellular superoxide dismutase in vitro, *Free Radical Biol. Med.* **13:**205–210.
29. Karlsson, K., and Marklund, S. L., 1987, Heparin-induced release of extracellular-superoxide dismutase to human blood plasma, *Biochem. J.* **242:**55–59.
30. Silverman, D. J., and Santucci, L. A., 1990, A potential protective role for thiols against cell injury caused by *Rickettsia rickettsii*, *Ann. N.Y. Accad. Sci.* **590:**111–117.
31. Puri, R. N., and Meister, A., 1983, Transport of glutathione, as gamma-glutamylcysteinylglycyl ester, into liver and kidney, *Proc. Natl. Acad. Sci. USA* **80:**5258–5260.
32. Devamanohaoran, P. S., Santucci, L. A., Hong, J. E., Tian, X., and Silverman, D. J., 1994,

Infection of human endothelial cells by *Rickettsia rickettsii* causes a significant reduction in the levels of key enzymes involved in protection against oxidative injury, *Infect. Immun.* **62:**2619–2621.

33. Blum, J., and Fridovich, I., 1985, Inactivation of glutathione peroxidase by superoxide radical, *Arch. Biochem. Biophys.* **240:**500–508.

34. Kono, Y., and Fridovich, I., 1982, Superoxide radical inhibits catalase, *J. Biol. Chem.* **257:**5751–5754.

35. Oliver, C. N., Levine, R. L., and Stadtman, E. R., 1987, A role of mixed-function oxidation reactions in the accumulation of altered enzyme forms during aging, *J. Am. Geriatr. Soc.* **35:**947–956.

36. Winkler, H. H., and Miller, E. T., 1981, Immediate cytotoxicity and phospholipase A: Role of phospholipase in the interaction of *R. prowazekii* and L-cells, in: *Proceedings of the Rocky Mountain Laboratory. Conference on Rickettsiae and Rickettsial Diseases* (W. Burgdorfer and R. L. Anacker, eds.), Academic Press, New York, pp. 327–333.

37. Winkler, H. H., and Daughtery, R. M., 1989, Phospholipase A activity associated with the growth of *Rickettsia prowazekii* in L-929 cells, *Infect. Immun.* **57:**36–40.

38. Walker, D. H., Firth, W. T., Ballard, J. G., and Hegarty, B. C., 1983, Role of phospholipase-associated penetration mechanism in cell injury by *Rickettsia rickettsii*, *Infect. Immun.* **40:**840–842.

39. Silverman, D. J., Santucci, L. A., Meyers, N., and Sekeyova, Z., 1992, Penetration of host cells by *Rickettsia rickettsii* appears to be mediated by a phospholipase of rickettsial origin, *Infect. Immun.* **60:**2733–2740.

40. Manor, E., Carbonetti, N. H., and Silverman, D. J., 1994, *Rickettsia rickettsii* has proteins with cross-reacting epitopes to eukaryotic phospholipase A_2 and phospholipase C, *Microb. Pathogen.* **17:**99–109.

41. Ryan, U. S., 1987, Endothelial cell activation responses, in: *Pulmonary Endothelium in Health and Disease* (U. S. Ryan, ed.), Dekker, New York, pp. 3–33.

42. Ryan, U. S., and Vann, J. W., 1988, Endothelial cells: A source and target of oxidant damage, in: *Oxygen Radicals in Biology and Medicine*, Volume 49 (M. E. Simic, K. A. Taylor, J. F. Ward, and C. von Sonntag, eds.), Plenum Press, New York, pp. 963–968.

43. Babior, G. L., Rosin, R. E., McMurrick, B. J., Peters, W. A., and Babior, B. M., 1981, Arrangement of the respiratory burst oxidase in the plasma membrane of the neutrophil, *J. Clin. Invest.* **67:**1724–1729.

44. Dewald, B., Baggiolini, M., Curnette, J. T., and Babior, B. M., 1979, Subcellular localization of the superoxide forming enzyme in human neutrophils, *J. Clin. Invest.* **63:**21–29.

45. Bromberg, Y., and Pick, E., 1985, Activation of NADPH-dependent superoxide production in a cell-free system by sodium dodecyl sulfate, *J. Biol. Chem.* **260:**13539–13542.

46. Curnutte, J. T., Badwey, J. A., Robinson, J. M., Karnovsky, M. J., and Karnovsky, M. L., 1984, Studies on the mechanism of superoxide release from human neutrophils stimulated with arachidonate, *J. Biol. Chem.* **259:**11851–11856.

47. Curnutte, J. T., 1985, Activation of human neutrophil nicotinamide adenine dinucleotide phosphate, reduced oxidase by arachidonic acid in a cell-free system, *J. Clin. Invest.* **75:**1740–1743.

48. Patriarca, P., Zatti, M., Cramer, P., and Rossi, F., 1970, Stimulation of the respiration of polymorphonuclear leukocytes by phospholipase C, *Life Sci.* **9:**841–844.

49. Tauber, A. I., 1985, Phagocyte NADPH-oxidase, in: *CRC Handbook of Methods for Oxygen Radical Research* (R. A. Greenwald, ed.), CRC Press, Boca Raton, pp. 25–30.

50. Dan, P., Nitzan, D. W., Dagan, A., Ginsburg, I., and Yedgar, S., 1996, H_2O_2 renders cells accessible to lysis by exogenous phospholipase A_2: A novel mechanism for cell damage in inflammatory processes, *FEBS Let.* **383:**75–78.

7

Intracellular Development of *Coxiella burnetii*

ROBERT A. HEINZEN

1. INTRODUCTION

The causative agent of human Q (query) fever, *Coxiella burnetii*, is a bacterial obligate intracellular parasite.[1] The pathogen has a worldwide distribution with a broad host range that includes humans, arthropods, fish, birds, rodents, and livestock.[2] In most wild or domestic animals, *C. burnetii* does not appear to cause overt disease. However, in goats and sheep it can cause abortions.[1] Humans become infected primarily by inhaling contaminated aerosols generated by domestic livestock operations.[2] In humans, the disease usually manifests as an acute flulike illness (Q fever) with symptoms that include high fever, severe preorbital headache, and hepatitis.[1] This acute form of the disease is treatable with antibiotics and is generally self-limiting. However, in rare instances chronic cases can occur that usually manifest as endocarditis. Q fever endocarditis is refractory to antibiotics and results in significant mortality.[3]

Although *C. burnetii* is included within the *Rickettsieae*, 16S rRNA sequences reveal that the organism is phylogenetically unrelated to the other genera within this tribe, namely, the *Rickettsia* and *Rochalimaea*.[4] *C. burnetii* is most closely related to the facultative intracellular bacterium, *Legionella pneumophila*.[4] Many phenotypic differences are observed between *C. burnetii* and other rickettsiae. For example, *C. burnetii* proliferates intracellularly within a membrane-bound vacuole with lysosomal characteristics[5-8] whereas members of the genus *Rickettsia* rapidly escape the

ROBERT A. HEINZEN • Department of Molecular Biology, University of Wyoming, Laramie, Wyoming 82071-3944.

Rickettsial Infection and Immunity, edited by Anderson *et al.* Plenum Press, New York, 1997.

early endosome and replicate within the host cell cytoplasm.[1] A hallmark that further distinguishes *C. burnetii* from other rickettsiae is its impressive long-lived extracellular stability and insensitivity to physical and chemical disruption.[9] The heat resistance of the pathogen is shown by survival in milk to temperatures of 63°C for 30 min.[9] The organism survives in dried guinea pig blood or in a 10% salt solution for at least 180 days at room temperature.[9] Resistances to 1% phenol or 0.5% formalin for 24 hr have also been reported.[2] *C. burnetii* remains highly infectious even after sonication of purified organisms in distilled water for over 30 min at 4°C (R. A. Heinzen and T. Hackstadt, unpublished data, 1996).

2. HOST–PARASITE INTERACTIONS

C. burnetii proliferates within a wide variety of epithelial, fibroblast, and macrophagelike cell lines.[1] The organism plays a passive role in adherence and entry into host cells. *C. burnetii* that are inactivated by heat or glutaraldehyde are endocytosed via a microfilament-dependent process at rates equal to those observed for viable bacteria.[10] The early parasite-containing phagosome proceeds through the endocytic pathway eventually acidifying to a pH of approximately 4.8.[7,11] *C. burnetii* has an absolute requirement for this moderately acidic pH to activate metabolism and growth *in vivo*, as well as metabolic activation *in vitro*.[12] The *C. burnetii*-containing vacuole fuses with lysosomes as demonstrated by the colocalization of the lysosomal enzymes 5'-nucleotidase,[5] acid phosphatase,[5–8] cathepsin D,[8] and two predominant lysosomal glycoproteins.[8] The organism undergoes luxurious growth within this vacuole, despite the presence of factors normally considered bactericidal. These can include reactive oxygen species generated during the phagocyte oxidative burst, and antimicrobial agents present in lysosomes such as acid hydrolases and defensins.[13] The *C. burnetii*-containing phagosome may be somewhat atypical as ingestion of the agent by macrophagelike cell lines results in a greatly diminished respiratory burst with little production of superoxide anion.[14] This phenomenon has also been observed during phagocytosis of *Leishmania*, *Histoplasma*, and some pathogenic *Mycobacteria*.[13]

3. SMALL-CELL AND LARGE-CELL VARIANTS

The remarkable extracellular stability of *C. burnetii* has been attributed to the existence of a small, resistant cell form that is part of a poorly defined developmental cycle.[15,16] Davis and Cox[17] first described the pleomorphic nature and filterability of *C. burnetii*. Using light microscopy, they characterized round particles that were filterable through 0.4-μm pores and fully infectious for guinea pigs.

Kordová[18,19] later determined that filterable forms are antigenically distinct from "mature" forms and suggested that they undergo a maturation process en route to typical *Coxiella*. It was from these studies that the concept of a *C. burnetii* developmental cycle was first introduced. In support of this hypothesis was a subsequent investigation[20] demonstrating that filterable forms undergo stepwise antigenic changes on infection of tissue culture cells. Early transmission electron microscopy (TEM) studies allowed clearer visualization of *C. burnetii* pleomorphism with resolution of some ultrastructure.[21-23] Nermut *et al.*[24] further differentiated cell types by demonstrating that large cells, but not small cells, are permeable to the negative stain, phosphotungstic acid. Independent groups subsequently demonstrated that small- and large-cell populations of *C. burnetii* can be separated on the basis of differences in buoyant density during equilibrium density centrifugation.[25-27] Two opposing theories addressing the formation of small- and large-cell forms were presented. One suggested that they represent morphological forms that occur during a *C. burnetii* developmental cycle.[25,27] The other suggested that small cells might arise from large cells as a result of phagolysosomal degradation.[25,27] Studies on the infectivity of large and small cells discounted the later theory.[25]

For the past 15 years, the vast majority of research on *C. burnetii* development has been conducted by McCaul and co-workers.[15,16,28-32] Employing more sophisticated fixation and staining procedures in conjunction with TEM, they conducted careful ultrastructural examinations of *C. burnetii* purified from infected yolk sacs, and organisms residing within infected tissue culture cells. Evidence obtained from a study conducted in 1981[15] prompted McCaul and Williams to propose a model for *C. burnetii* development that includes both vegetative and sporogenic morphological differentiations (Fig. 1). In the hypothesized developmental cycle, metabolically dormant small cells [designated small-cell variants (SCV)], or potentially sporelike forms, are phagocytosed by a eukaryotic host cell and sequestered in a phagolysosome. Exposure to low intraphagolysosomal pH and, perhaps, enzyme systems and/or nutrient sources present in the vacuole, triggers metabolism of dormant SCV that subsequently undergo vegetative differentiation into more metabolically active large cells. Large cells [designated large-cell variants (LCV)], on sensing the demise of the host, initiate a process resembling sporogenesis. "Endospores" are then liberated on lysis of LCV. The authors speculated that SCV or "spores" are extracellular survival forms.

Although the issue of sporogenesis remains to be resolved, TEM studies have clearly demonstrated a continuum of vegetative morphological forms, including SCV and LCV, that occur during *C. burnetii* development.[15,16,21-38] The SCV population has been further subdivided into two groups on the basis of pressure sensitivity. Cells that are insensitive to breakage in a French press at 20,000 lb/in.[2] are referred to as small dense cells (SDC).[30] For the purpose of this review, the distinction between SCV and SDC will not be made. Rather, the focus will be

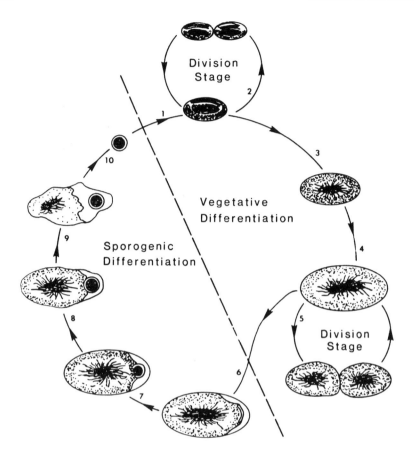

FIGURE 1. Putative *Coxiella burnetii* developmental cycle as proposed by McCaul and Williams.[15] Metabolically dormant "spores" or small-cell variants (SCV) are endocytosed by the host cell and sequestered in a phagocytic vacuole that eventually fuses with lysosomes. Low pH and nutrient sources activate metabolism and promote vegetative differentiation of SCV to the more metabolically active large-cell variant (LCV) (stages 1–5). The demise of the host may trigger the production of particles resembling endospores by a subpopulation of LCV (stages 6–10). Lysis of the host cell releases all cell forms with only the SCV (and possibly a sporelike form) able to persist extracellularly. (Reprinted from McCaul, T. F., and Williams, J. C., 1981, *J. Bacteriol.* **147:**1063–1076. With permission.)

on the two cell populations that can be purified to relative homogeneity by equilibrium centrifugation, namely, the SCV (containing SDC) and LCV. As discussed below, these are the cell populations that are the most thoroughly characterized in terms of antigenicity, infectivity, ultrastructure, metabolic activity, physical properties, and specific protein constituents.

3.1. Purification

SCV and LCV of phase I [full-length lipopolysaccharide (LPS)[1]] virulent *C. burnetii* can be purified on the basis of different buoyant densities utilizing centrifugation to equilibrium in cesium chloride (CsCl),[25,37,38] or density gradients of sucrose or Renografin.[25–27] There is a direct correlation between the buoyant density of LCV and SCV, and the osmotic pressure exerted by the centrifugation medium.[25] Of the three mediums, Tris-buffered 32% CsCl is most effective in partitioning the two cell types.[25] When purified organisms are subjected to this procedure, two bands are visible: a thick, upper band containing SCV (density = 1.280 g/ml) and a thin, lower band containing LCV (density = 1.324 g/ml).[25,37,38] The permeability of the LCV, but not the SCV, to the CsCl salt solution is thought to result in the LCV partitioning as the denser cell form.[25] The yield of SCV is consistently more than 5 times greater than that of the LCV, on a milligram dry weight basis. This result is probably related to the fragility of the latter cell type.[38] SCV and LCV preparations are generally at least 95% homogeneous based on size and morphological criteria[37,38] (Fig. 2). Although CsCl is effective in separating SCV and LCV, it is osmotically unfavorable and purified *Coxiella* demonstrate reduced infectivity and greatly diminished metabolic capabilities *in vitro* (Refs. 25, 38; R. A. Heinzen and T. Hackstadt, unpublished data, 1996). Nevertheless, these preparations are quite useful for ultrastructural studies and as a source of SCV and LCV protein. The hydrophobicity of *C. burnetii* in phase II (truncated LPS[1]) precludes separation of SCV and LCV in 32% CsCl because of severe clumping of organisms (R. A. Heinzen and T. Hackstadt, unpublished data, 1996). However, preparations that are highly enriched for SCV of both phase I and II can be obtained by procedures that exploit the sensitivity of LCV to physical disruption (Refs. 28, 30, 33; R. A. Heinzen and T. Hackstadt, unpublished data, 1996).

3.2. Ultrastructure

McCaul and co-workers[15,16] have defined in exquisite detail the ultrastructural differences between LCV and SCV. The SCV is typically between 0.2 and 0.5 μm in size, rod shaped, and very compact (Fig. 2A). The visible periplasmic space is replaced with an electron-dense region bounded by the outer and cytoplasmic membranes. The most distinctive ultrastructural characteristic of the SCV is the electron-dense condensed chromatin. This structure was commented on in earlier studies[21–24] and was referred to as a "dense central body"[23] or "compact central mass"[22] and correctly interpreted to contain nucleic acid.[21,23,24] The SCV also harbors a complex system of internal membranes, arranged in whorls, that may be continuous with the cytoplasmic membrane.[15,16,23]

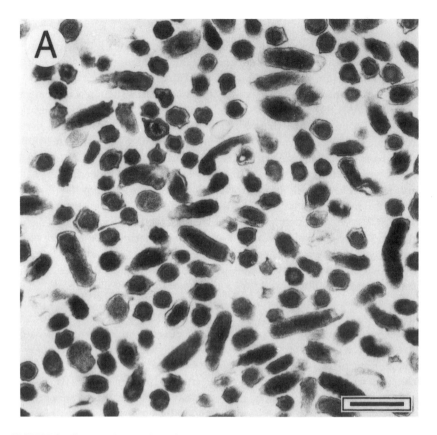

FIGURE 2. Electron micrographs of *C. burnetii* phase I SCV and LCV separated by cesium chloride equilibrium density centrifugation. (A) Electron micrograph of purified SCV. Note size and characteristic condensed chromatin. (B) Electron micrograph of purified LCV. Note size and characteristic dispersed chromatin. Bars = 0.5 μm.

The LCV can reach a length exceeding 1.0 μm and is similar to typical gram-negative bacteria in possessing a clearly distinguishable outer membrane, periplasmic space, and cytoplasmic membrane (Fig. 2B). The LCV is more pleomorphic than the SCV with a thinner cell wall and a very dispersed nucleoid. As depicted in Fig. 3 and described by others,[15,16] both SCV and LCV are capable of dividing by binary fission.

3.3. Infectivity

SCV and LCV are infectious for *in vitro* and *in vivo* models. Wiebe *et al.*[25] demonstrated that CsCl-separated SCV and LCV are infectious for chick yolk sac

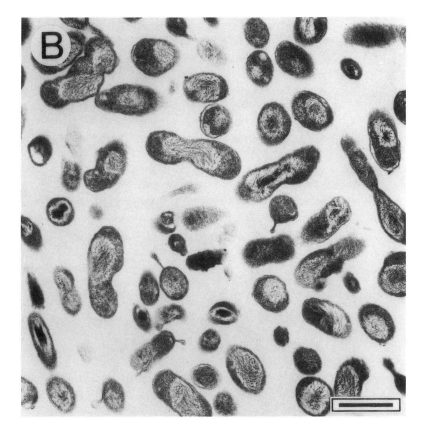

FIGURE 2. (*Continued.*)

cells and lethal to chick embryos. Infection by SCV or LCV eventually results in phagolysosomes harboring a mixture of cell types.[25] This observation discounted the notion that LCV are simply noninfectious, degenerative SCV.[25,27] A caveat to this conclusion is that centrifugation techniques do not purify cell types to homogeneity. Therefore, infection could be the result of contaminating SCV in LCV preparations. However, when Vero cells are infected with a mixture of phase I or phase II SCV and LCV, both cell types are observed within membrane-bound vacuoles as early as 2 hr postinfection (Fig. 4). This observation demonstrates that both cell forms are internalized and presumably capable of proceeding through the infectious cycle. It is unlikely that significant differentiation has occurred during the 2-hr infection as the organism has a very slow growth rate. Although the generation time of *C. burnetii* has yet to be determined, growth of the organism is tediously slow as evidenced by the 16-day incubation period required to produce visible plaques on primary chick embryo fibroblasts.[1]

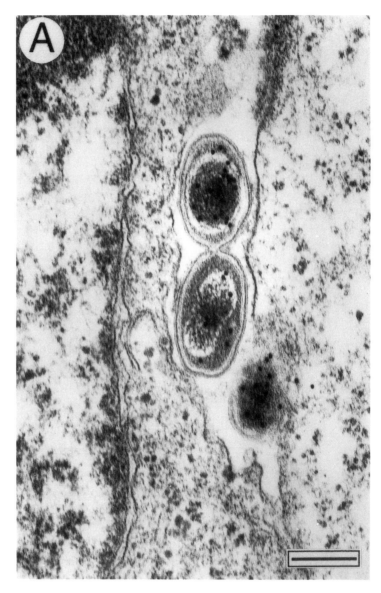

FIGURE 3. Electron micrographs of Vero cells infected with *C. burnetii*, phase II, showing (A) SCV and (B) LCV in the process of dividing by binary fission. The SCV in (A) are immunogold labeled with antiserum against the SCV-specific protein, ScvA.[38] Bars = 0.2 μm.

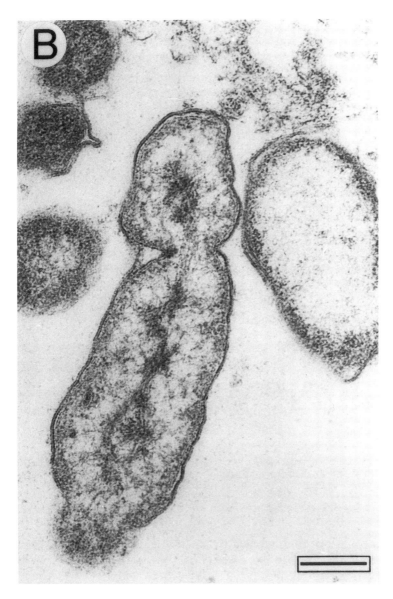

FIGURE 3. (*Continued.*)

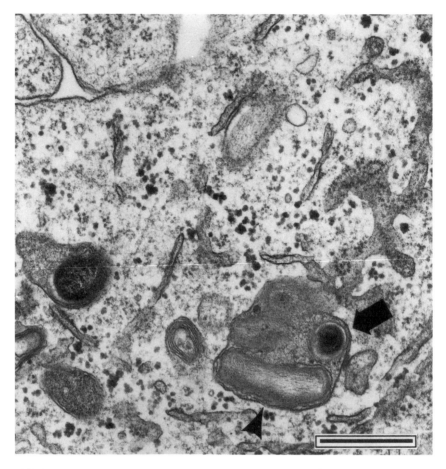

FIGURE 4. Electron micrograph of a Vero cell infected for 2 hr with a mixture of SCV and LCV of *C. burnetii*, phase II. A vacuole harboring both an SCV (arrowhead) and LCV (dagger) is evident. Bar = 0.5 μm.

The fact that LCV are infectious *in vitro* may have little relevance to natural transmission and infection because LCV likely do not persist extracellularly in an infectious form for extended periods.[16,28]

3.4. Resistance and Metabolic Properties

When metabolically activated in axenic medium,[12] the SCV is much less active than the LCV. This characteristic and the enhanced resistance of the SCV to disruption were both illustrated in a study by McCaul *et al.*[28] Purified *C. burnetii*, containing approximately 35% LCV, was subjected to osmotic shock in distilled water, incubation at 45°C for 2 hr, sonication for 45 min at 4°C, and centrifugation through 40–70% sucrose gradients. The resultant fraction, containing SCV with no intact LCV, exhibited a 14- and 43-fold reduction in the rate of metabolism of glutamate and glucose, respectively, when compared with the starting material.

The resistance of the SCV to osmotic shock and sonication correlates with an impermeability to negative stains, such as phosphotungstic acid[16,24,28,30] or ammonium molybdate. Renografin-purified *C. burnetii* that have been subjected to osmotic shock in distilled water for 2 hr at 37°C, followed by 8 min of sonication, contain mostly intact SCV and LCV cell wall material with an occasional intact LCV. When this preparation is negatively stained with 1% ammonium molybdate, SCV are impermeable to the stain. Conversely, LCV and LCV cell wall material are stained (Fig. 5).

Three reports appear to support the hypothesis that the *C. burnetii* cell cycle contains cell types with different metabolic capabilities.[39–41] Each investigation examined two cell populations derived from persistently infected BHK-21 (baby hamster kidney) cells. Naturally released cells (NRC) are *Coxiella* that accumulate in tissue culture medium over a period of 12 to 16 hr as a consequence of release from infected cells by either exocytosis or lytic events. Mechanically released cells (MRC) are *Coxiella* that are obtained by short cycles of vortexing infected fibroblasts. After rapid purification of the organisms from host material and metabolic activation in a low-pH, axenic buffer, the two cell populations exhibit distinctly different properties with regard to transport and incorporation of substrates. For example, MRC transport and incorporate much higher levels of uridine and leucine when compared to NRC.[39–41] SDS-PAGE profiles of labeled polypeptides extracted at different times during the incubation reveal notable differences between NRC and MRC in the type and amount of proteins produced and the time during incubation at which certain proteins are synthesized.[39–41] NRC also export much more protein into the surrounding medium than MRC.[40,41] It was proposed that the biosynthetic capabilities observed with NRC are those that an extracellular transit form of *C. burnetii* might exhibit on initially encountering the phagolysosomal environment.[39–41] Conversely, it was suggested that the metabolism witnessed with MRC may more closely resemble that of rickettsiae actively growing and dividing within the phagolysosome.[39–41] It is tempting to speculate that "secretory competent" NRC synthesize and release proteins that modify the early phagolysosomal vacuole and make it more amenable for growth.[14,41,42] It could not be ascertained whether LCV or SCV are largely responsible for the differential metabolic properties of NRC and MRC as TEM revealed that both cell populations contain organisms in various stages of differentiation.[31]

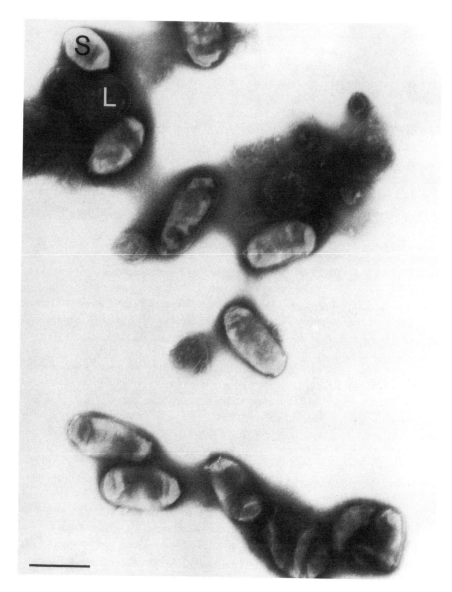

FIGURE 5. Electron micrograph of purified *C. burnetii*, phase II, negatively stained with 1% ammonium molybdate. Cells were osmotically shocked in distilled water for 2 hr followed by sonication for 8 min prior to staining. Representative SCV and LCV are labeled S and L, respectively. Bar = 0.5 μm.

3.5. Similarities with *Chlamydia*

Chlamydia are obligate intracellular bacterial parasites of eukaryotic cells that undergo a biphasic life cycle within membrane-bound vacuoles that is superficially similar to that of *C. burnetii*.[43] Although *Chlamydia* and *Coxiella* are phylogenetically distant, their respective morphological forms exhibit some similar biological and ultrastructural properties.[16,43]

The *C. burnetii* SCV and the chlamydial elementary body (EB) are both metabolically dormant, resistant small cell types that are adapted for extracellular survival.[16,43] When endocytosed and presented with the correct intravacuolar environment, both resistant cell types differentiate into large, more metabolically active cell types, namely, the LCV of *C. burnetii* and the reticulate body (RB) of chlamydia.[16,43] Like LCV, the RB is fragile and unstable outside the host.[16,43] Both LCV and RB divide by binary fission throughout the remainder of the infectious cycle.[16,43] At some point during the chlamydial infectious cycle (about 18 hr postinfection for *Chlamydia trachomatis*, serovar L2), development becomes asynchronous as some RB begin to differentiate and condense back to EB which accumulate within the vacuole until host cell lysis occurs.[43]

Notwithstanding these similar biological properties, there are some notable differences. For example, SCV undergo binary fission[15,16] (Fig. 3) whereas EB do not divide.[43] Furthermore, LCV, but not RB, are infectious[25] (Fig. 4). Although SCV-to-LCV morphogenesis mimics that observed during EB-to-RB differentiation, there is currently no evidence of a "condensing" LCV-to-SCV transition that emulates chlamydial RB-to-EB differentiation.[15] However, the lack of a synchronous *C. burnetii* infection model precludes eliminating such an event.

The ultrastructural parallels between *C. burnetii* and the chlamydiae have been commented on by several groups.[22,25,35] The most obvious similarity is the condensed chromatin of both EB[43] and SCV.[16] The chromatin is dispersed and filamentous in both RB[43] and LCV[16]. As discussed later in this chapter, the condensed chromatin of both *Chlamydia* EB and *Coxiella* SCV contain eukaryotic lysine-rich histone H1 homologues.[37,44,45]

3.6. SCV- and LCV-Specific Components

Cells undergoing differentiation display developmentally regulated synthesis of subsets of proteins. Temporal and/or preferential transcription of genes is necessary for progression to a different cell morphology. This phenomenon has been well studied in bacterial systems that include the spore-forming bacilli[46] and streptomycetes,[47] and the social behavior displayed by myxobacteria.[48]

There is a paucity of information on developmentally regulated protein synthesis involved in *C. burnetii* morphogenesis. However, the morphological differences between SCV and LCV do correlate with differences in protein composi-

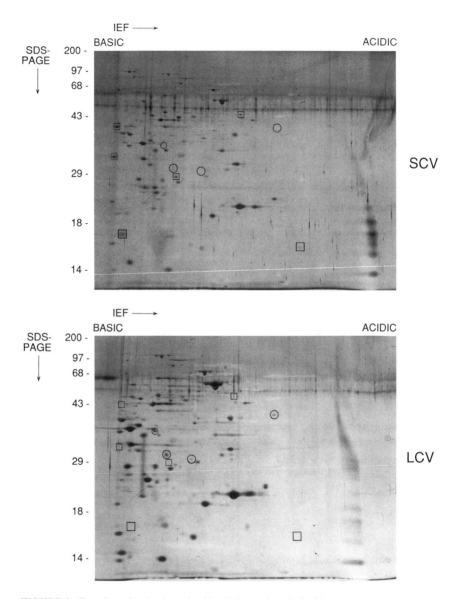

FIGURE 6. Two-dimensional polyacrylamide gel electrophoresis (PAGE) and silver staining of SCV and LCV lysates. Purified cell forms were lysed in sample buffer and subjected to isoelectric focusing and sodium dodecyl sulfate-12.5% PAGE followed by silver staining. A number of differentially expressed proteins are observed with the SCV (squares) and LCV (circles). Molecular mass is expressed in kilodaltons.

tion.[16,29,30,37,38] As depicted in Fig. 6, two-dimensional polyacrylamide gel electrophoresis and silver staining of *C. burnetii* cell lysates reveal a number of proteins and/or carbohydrates specific to each cell form. SCV-specific components predominate at lower M_rs whereas LCV-specific proteins are more prevalent at higher M_rs.[38] Biochemical analysis also reveals that LCV cell walls contain 14% more protein, 2.7-fold less peptidoglycan, and have a different amino acid composition from SCV.[33] As discussed below, modern molecular techniques such as peptide sequencing in combination with gene cloning are now allowing identification and characterization of proteins differentially synthesized by SCV and LCV.

Histone H1 Homologue

Compact nucleoid structures of prokaryotes contain basic DNA binding proteins that serve structural roles and effect gene expression by introducing topological changes in DNA.[49] Using a filter screening technique for DNA binding proteins referred to as "Southwestern" blotting, two histone H1 homologues from *Chlamydia trachomatis* have been identified and their respective genes cloned.[44,45] These proteins are associated with the condensed nucleoid of the chlamydial EB.[44,45] Because the condensed nucleoid of the *C. burnetii* SCV is similar in appearance to that observed for chlamydial EB, it was surmised that this structure may also contain histonelike proteins. Using a similar methodology, two *C. burnetii* DNA binding proteins have been identified termed Hq1 (20 kDa) and Hq2 (14 kDa).[37] Hq1 is highly enriched in purified SCV whereas Hq2 is present in approximately equal amounts in SCV and LCV. Heparin agarose affinity chromatography allowed purification of Hq1 on the basis of its positive charge. The Hq1 N-terminal amino acid sequence was subsequently deduced and degenerate oligonucleotide probes corresponding to this sequence were used to clone the encoding gene, designated *hcbA*. The derived amino acid sequence of Hq1 predicts a protein of 119 amino acids with a molecular mass of 13,183 Da and a highly basic isoelectric point (pI) of 13.1. Hq1 exhibits 34 and 26% identity with eukaryotic histone H1 and *C. trachomatis* Hc1, respectively. This identity is primarily reflective of the high lysine content of these proteins (Table I). The association of Hq1 with the SCV implies that its synthesis is developmentally regulated. The ability of Hq1 to bind DNA *in vitro* suggests that it may play a pivotal role in control of gene expression in *C. burnetii*.

ScvA

The high isoelectric point of histonelike proteins, such as Hq1, was exploited to purify another basic protein associated with the SCV chromatin termed ScvA (*small cell variant protein A*).[38] When *C. burnetii* was solubilized in the nonionic dializable detergent octylglucopyranoside (OGP) and subjected to preparative

TABLE I
Amino Acid Composition of Eukaryotic Histone H1 from the Painted Sea Urchin
(*Lytechinus pictus*) and the Corresponding Homologues of *Chlamydia*
***trachomatis* (Hc1 and Hc2) and *C. burnetii* (Hq1)**

H1 *Lytechinus pictus* 209 aa molecular wt: 21,603 pI: 11.8			Hc1 *Chlamydia trachomatis* 125 aa molecular wt: 13,682 pI: 11.4			Hc2 *Chlamydia trachomatis* 222 aa molecular wt: 23,663 pI: 13.7			Hq1 *Coxiella burnetii* 117 aa molecular wt: 13,181 pI: 13.1		
aa	#	mol %	aa	#	mol %	aa	#	mol %	aa	#	mol %
A Ala	62	29.67	A Ala	23	18.40	A Ala	52	23.42	A Ala	22	18.80
C Cys	0	0.00	C Cys	2	1.60	C Cys	3	1.35	C Cys	0	0.00
D Asp	1	0.48	D Asp	2	1.60	D Asp	0	0.00	D Asp	2	1.71
E Glu	5	2.39	E Glu	6	4.80	E Glu	0	0.00	E Glu	2	1.71
F Phe	2	0.96	F Phe	0	0.00	F Phe	0	0.00	F Phe	0	0.00
G Gly	9	4.31	G Gly	2	1.60	G Gly	4	1.80	G Gly	6	5.13
H His	1	0.48	H His	0	0.00	H His	3	1.35	H His	0	0.00
I Ile	5	2.39	I Ile	3	2.40	I Ile	0	0.00	I Ile	0	0.00
K Lys	66	31.58	K Lys	36	28.80	K Lys	56	25.23	K Lys	29	24.79
L Leu	6	2.87	L Leu	7	5.60	L Leu	3	1.35	L Leu	5	4.27
M Met	2	0.96	M Met	4	3.20	M Met	4	1.80	M Met	2	1.71
N Asn	4	1.91	N Asn	2	1.60	N Asn	2	0.90	N Asn	1	0.85
P Pro	12	5.74	P Pro	3	2.40	P Pro	9	4.05	P Pro	3	2.56
Q Gln	6	2.87	Q Gln	3	2.40	Q Gln	3	1.35	Q Gln	3	2.56
R Arg	5	2.39	R Arg	6	4.80	R Arg	21	9.46	R Arg	24	20.51
S Ser	6	2.87	S Ser	5	4.00	S Ser	8	3.60	S Ser	8	6.84
T Thr	7	3.35	T Thr	14	11.20	T Thr	20	9.01	T Thr	7	5.98
V Val	8	3.83	V Val	6	4.80	V Val	32	14.41	V Val	3	2.56
W Trp	0	0.00	W Trp	0	0.00	W Trp	1	0.45	W Trp	0	0.00
Y Tyr	2	0.96	Y Tyr	1	0.80	Y Tyr	0	0.00	Y Tyr	0	0.00
B Asx	0	0.00	B Asx	0	0.00	B Asx	0	0.00	B Asx	0	0.00
Z Glx	0	0.00	Z Glx	0	0.00	Z Glx	0	0.00	Z Glx	0	0.00

isoelectric focusing, a small protein (approximately 4 kDa) separated in the highest pI fraction (approximately pH 11).[38] Degenerate deoxyoligonucleotides corresponding to the N-terminal sequence of this protein were used to recover a cosmid clone from a *C. burnetii* genomic library. Nucleotide sequencing of insert DNA revealed an open reading frame designated *scvA* with coding potential for the 30-amino-acid protein MERQNVQQQRGKDQRPQRPGASNPRRPN-QR (ScvA) with a predicted mass and pI of 3610 Da and 12.6, respectively. The basic nature of ScvA is conferred primarily by the high arginine content (23%). ScvA is also 23% glutamine and 13% proline. Immunoblot analysis with ScvA antiserum revealed that ScvA is present only in lysates of purified SCV, and not LCV. These data correlate with results obtained by immunogold electron microscopy that demonstrate an abundance of label associated with purified SCV, primarily confined to the central core of condensed chromatin (Fig. 7A). Only an occasional gold particle is found in sections of purified LCV, generally associated with a contaminating SCV (Fig. 7B). Unlike Hq1 and Hq2, immunoblot and DNA filter hybridizations indicate that a homologue of *C. burnetii* ScvA is not present in chlamydia.[38]

The basic nature and affinity of ScvA for chromatin coincides with the ability of this protein to bind DNA *in vitro*.[38] At a molar ratio approximating 10 ScvA molecules to 1 double-stranded supercoiled plasmid molecule, a protein:DNA complex forms that retards the mobility of plasmid DNA through a 0.8% agarose gel. At a ratio approximating 50:1, a protein:DNA complex forms that does not enter the gel.

Although assigning roles for ScvA and Hq1 is speculative, they may serve a structural role in the formation of the SCV condensed nucleoid. This assumption is based on homology (Hq1), *in vitro* interactions with DNA (Hq1 and ScvA), and the predominance of labeling of the chromatin by immunogold electron microscopy (ScvA). Binding of DNA by ScvA and Hq1 is probably not DNA sequence specific as irrelevant plasmid DNA was employed as substrate in *in vitro* assays.[37,38] Binding of *C. burnetii* genomic DNA *in vivo* by one or both of these proteins could potentially serve a protective role, or induce topological changes that alter gene expression.[49,50] It is unknown if ScvA cooperatively associates in some fashion with Hq1.

It is enticing to draw an analogy between ScvA and the small acid-soluble proteins (SASP) of *Bacillus* spores.[51] This family of low-molecular-weight (5–11 kDa) proteins nonspecifically bind double-stranded DNA[52] and protect spore DNA from insult by ultraviolet light[51] and heat[53]. They comprise 10–20% of the protein content of spores and are rapidly degraded by SASP-specific proteases during spore germination with the released amino acids used for biosynthesis.[51] ScvA also appears to be degraded by recently internalized SCV where a 63% reduction in the protein, when compared with the starting inoculum, is observed by immunoblotting at 12 hr postinfection.[38] Hackstadt and Williams[12] demonstrated that oxygen is consumed by substrate-starved organisms that are metabolically activated at low pH in axenic medium, implying depletion of an endogenous reserve. Whether ScvA comprises part of this reserve is unknown. ScvA is 23% glutamine, an amino acid that is readily converted to glutamate, an important energy and biosynthetic source of *C. burnetii*.[12] Therefore, it is conceivable that ScvA, like SASPs, may act as a nutritional source during the early stages of metabolic activation of the SCV, and that its degradation correlates with outgrowth of SCV to LCV.

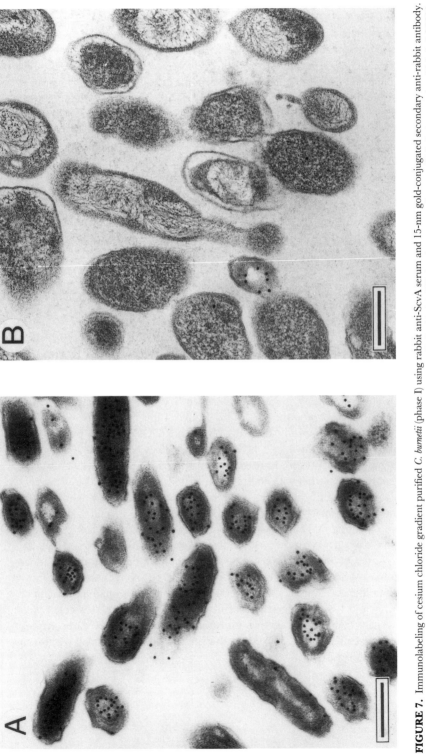

FIGURE 7. Immunolabeling of cesium chloride gradient purified *C. burnetii* (phase I) using rabbit anti-ScvA serum and 15-nm gold-conjugated secondary anti-rabbit antibody. (A) Purified SCV showing an abundance of label associated with the condensed chromatin. (B) Purified LCV showing a near absence of label. An occasional labeled SCV is observed in LCV preparations. Bars = 0.2 μm.

29.5 kDa and LPS

McCaul and co-workers reported that monoclonal antibodies raised against a 29.5-kDa *C. burnetii* outer membrane protein termed P1[54] label LCV but not SCV.[16,30] The authors concluded that the epitopes recognized by these antibodies are buried in the SCV outer membrane and not accessible to antibody, or that the SCV does not synthesize this protein. The latter scenario is likely correct as P1 is similarly not found associated with the SCV when cell lysates are examined by Coomassie brilliant blue staining or immunoblot analysis of SDS-PAGE-separated proteins.[30] It has been reported that P1 is an acid-activatable porin[54] with N-terminal amino acid homology with *E. coli* OmpA (J. E. Samuel, personal communication). In *E. coli*, OmpA functions as a pore in allowing the nonspecific, slow diffusion of small solutes.[55] This raises the possibility that, unlike the SCV, the more metabolically active LCV requires this pore to obtain essential nutrients, such as amino acids, from the host. *C. burnetii* P1 is also highly immunogenic.[56] As alluded to by others (Ref. 30; J. E. Samuel, personal communication), the absence of this protein on the SCV may be a mechanism by which *C. burnetii* evades the host immune response and establishes chronic infection.

When examined by immunogold TEM with a monoclonal antibody against phase I LPS, SCV appear substantially enriched for LPS when compared to the LCV.[16,30] This introduced the possibility that LPS synthesis is developmentally regulated.[16,30] An alternative explanation suggested the deficiency of LPS on the LCV is the result of enzymatic degradation and/or sloughing of the molecule during infection.[16,30] LPS is known to be loosely associated and shed during infection by other intracellular bacteria, such as chlamydia.[57] There is not a total absence of LPS on the LCV as the molecule is detected on silver-stained gels and immunoblots of LCV cell walls.[30] Consequently, the distinction still needs to be made between selective synthesis of LPS by the SCV as opposed to enhanced shedding of LPS by the LCV during infection and/or purification.

4. SPORELIKE PARTICLES

There are obvious morphological and compositional changes coincident with developmental transition from SCV to LCV. Of a more controversial nature is the concept of sporogenesis in this bacterium. This alternative form of *C. burnetii* replication was originally forwarded by McCaul and Williams in 1981[15] and is based on morphological data gathered by TEM examination of purified cells (Fig. 1). The impetus for this proposal was the occasional appearance of an electron-dense, membrane-bound, polar body within LCV that morphologically appears to arise as a result of asymmetric septation. Because the term *endospore* implies specific physical and structural characteristics,[58] most of which are lacking in the

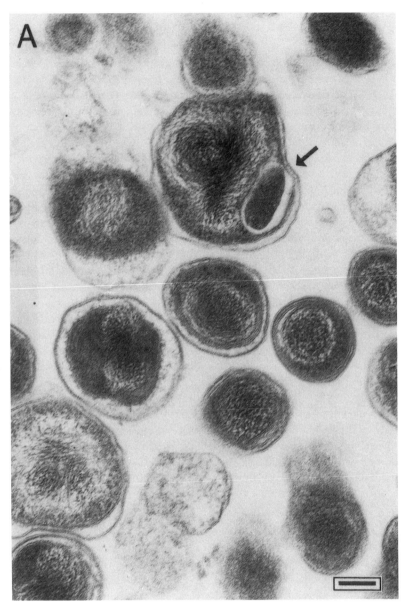

FIGURE 8. Electron micrographs of a *C. burnetti*, phase II, cell showing a typical sporelike particle (SLP) and comparison to electron-dense bodies of mycobacteria. (A) SLP in *C. burnetii* purified from an infected Vero cell culture incubated at room temperature for 28 days with no media changes prior harvest of microorganisms. (B) Electron-dense structure of *Mycobacterium marinum*. Bars = 0.1 μm. (B courtesy of Pam Small and Fred Hayes, Rocky Mountain Laboratories.)

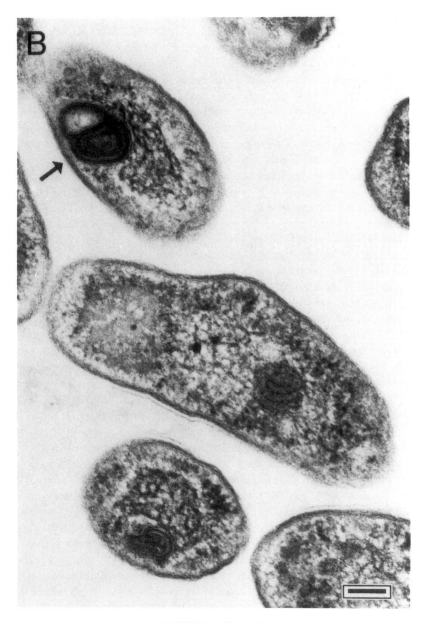

FIGURE 8. (*Continued.*)

postulated *C. burnetii* counterpart, this review will refer to these structure as *sporelike particles* (SLP), a term originally coined by Schaal *et al.*[36]

The ultrastructure of the SLP, and a model of postulated steps in its morphogenesis, have been reviewed in detail by McCaul.[16] SLP are approximately 130–170 nm in diameter and are bounded by a limiting membrane (Fig. 8A). The particle is observed only in those cells morphologically classified as LCV. Formation of SLP generally occurs in a polar location by a process that resembles asymmetric cell division. From the center outward the SLP is believed to be comprised of a dense core, a system of membranes, peptidoglycan, and an outer membrane. Some of the morphological features thought to occur during SLP formation are reminiscent of those observed during sporulation of *Bacillus*.[46]

Is the SLP a true endosphere in structure and function? Arguments based on the experimental evidence can be made both in favor of and against this proposition. Two recent reports appear to support the theory that the SLP is a spore. First, postembedding immunoelectron microscopy has revealed that the SLP contains DNA.[29] Second, the predicted product of a *C. burnetii* open reading frame (ORF 274) has been shown to be homologous to *Bacillus subtilis* SpoIIIE.[59] The *spoIIIE* gene product is a regulator of prespore-specific gene expression.[60] The function of the ORF 274 product in *C. burnetii* is unknown. The possibility exists that it may serve a regulatory role unrelated to sporogenesis. It would be interesting to determine whether the *C. burnetii* genome contains sequences homologous to *Bacillus* spore structural genes.

There are several observations that appear to contradict the idea of *C. burnetii* sporogenesis. For example, the biochemical spore-marker, dipicolinic acid, is not detected in *C. burnetii* and traditional spore stains have little affinity for the SLP.[15] Moreover, the resistance of *C. burnetii* to various physical stresses are intermediate those of vegetative cells and spores.[9] It has been proposed that the SLP functions as an extracellular survival form of *C. burnetii* and that these particles may play a role in the persistence of the agent observed in chronic infections.[15,16,32] Indeed, SLP have recently been detected in infected cardiac valves of patients with Q fever endocarditis.[32] However, there is no teleological reason for a "sporelike" stage of the developmental cycle as the physical properties of the SCV are sufficient to explain the extracellular stability of *C. burnetii*.[28]

A considerable impediment to the study of the SLP is an inability to obtain these particles in pure form. This would allow a detailed biochemical analysis, in addition to experiments designed to assess whether they are the infectious precursor of the SCV. *Bacillus* and other spore formers synthesize a number of proteins unique to the endospore, such as those that comprise the spore cortex and coat.[58] However, the specific antigenic character of the SLP remains elusive. Broadly specific cytochemical stains for polysaccharides, mucopolysaccharides, mucopolysaccharide–protein complexes, peptidoglycanlike substances, and nonpeptidoglycan protein polymers stain the outer surface of the SLP.[16] However, polyclonal antisera directed against *C. burnetii* (phase I) whole cells, delipidated whole cells, cell walls, and peptidoglycan–protein complexes, do not label the SLP by postembedding immunogold TEM.[16,30] Similarly, monoclonal antibodies against phase I LPS and the P1 protein do not label the SLP.[16,30] It was concluded from these studies that the antigenic composition of the SLP is distinct from that of the SCV and LCV. The SLP also does not label with ScvA antiserum[38]; consequently, an SLP-specific protein has yet to be identified.

If not a spore, then what are some alternate explanations for the SLP? Perhaps they represent mesosomal-type structures described for other prokaryotes.[61] Although their function is obscure, these membranous intrusions of the cytoplasmic membrane are known to interact with DNA. Mesosomal structures have been described in diverse bacteria including members of the genus *Rickettsia*[62] and *Mycobacteria*[63] (Fig. 8B). Interestingly, sporelike bodies termed coccal forms have also been reported in mycobacteria, but their function is unknown.[63] Another possibility is that the SLP represents a type of storage granule.[61] These are observed as inclusion bodies within the bacterial cytoplasm and are usually produced under conditions of nutrient deprivation. They serve to concentrate a variety of compounds such as polyphosphate, glycogen, and β-hydroxybutyrate and, depending on their composition, can appear as electron-dense objects by TEM. In contrast to the SLP, multiple storage granules can be observed in a single cell and they are not surrounded by membrane or encompassed by an atypical membrane. Perhaps the *C. burnetii* SLP is analogous to a bacterial cyst. Cysts produced in response to stress by the soil-inhabiting *Azotobacteriaceae* are filterable through a 0.45-μm filter, have no dipicolinic acid, demonstrate heat resistance intermediate between that of vegetative cells and spores, and are known to survive up to 24 years in dry soils.[64] Finally, one needs to consider the possibility that the SLP is an artifact of dehydration and fixation procedures used in sample preparation for TEM. Slight modifications of these steps have been shown to result in vastly different ultrastructural images.[65]

5. POTENTIAL ENVIRONMENTAL MODULATORS OF DEVELOPMENT

The environmental signals that govern *C. burnetii* morphogenesis are unknown. The two most obvious candidates are nutrient availability and pH. Because both SCV and LCV are infectious, and a typical late phagolysosome always contains a mix of cell forms, it is unlikely that development is a mechanism for survival against the normally bactericidal elements of lysosomes. Occasionally, LCV are witnessed within infected cells that appear in various stages of degradation,[5–7,31,34] but the significance of this observation is unclear. These may simply be "dead" LCV that are now more susceptible to digestion by lysosomal enzymes.

Alternatively, viable LCV may be inherently less resistant to digestion than SCV with some cell death occurring during the infectious cycle. Because both humoral and cell-mediated immune responses are elicited during *C. burnetii* infection,[54,56] some fragmentation of rickettsial components must occur, at least by antigen-presenting cells such as macrophages.

The resistance to lysosomal degradation is likely conferred by the cell surface of *C. burnetii*. The agent appears to have a typical gram-negative-type outer membrane.[54] However, the LPS is unusual in that the O-antigen contains some uncommon sugars and the core and lipid regions are comprised of biochemically unique components.[54] Given that phase II (rough LPS) *Coxiella* grow equally as well as phase I (smooth LPS) in tissue culture,[1] the putative resistance properties of LPS would have to be present in the lipid A-core region.

5.1. Nutrient Availability

A more probable regulator of *C. burnetii* development is the metabolic status of the host. Although not yet tested, the parasitic burden imposed by bacterial growth late in the infectious cycle probably inflicts nutritional stress on the host. Heavily infected host cells that are degenerating may reduce trafficking of nutrient-laden vesicles to the *C. burnetii*-containing vacuole.[8,54] This, in turn, may drive development of *C. burnetii* to a population dominated by the SCV, the cell type most likely to survive extracellularly. Depletion of carbon, nitrogen, or phosphorous sources is a known trigger of bacterial morphogenesis, such as *Bacillus* sporulation[46] and *Azotobacter* cyst formation,[64] and may similarly promote morphological differentiation of *C. burnetii*. An interesting observation that may be related to this hypothesis is that the percentage of LCV harboring SLP markedly increases in *C. burnetii*-infected cells maintained in culture for long periods (6 weeks or more) without media changes (Ref. 36; R. A. Heinzen and T. Hackstadt, unpublished data, 1996) (Fig. 8). Although it has yet to be demonstrated that the SLP is an infectious precursor of the SCV, increased production of these particles appears to be associated with nutritionally stressed host cells. SLP are rarely observed in optimally maintained cultures of infected cells (Ref. 36; R. A. Heinzen and T. Hackstadt, unpublished data, 1996).

5.2. pH

Small changes in external and internal pH are known to modulate bacterial gene expression.[66] As discussed earlier, *C. burnetii* has an absolute requirement for the moderately acidic pH found in the phagolysosome to activate its metabolism.[12] The pH-dependent activation of metabolism by whole cells in axenic medium has been partially explained by the resulting energization of membranes and substrate transporters.[12] Like more extreme acidophiles, *C. burnetii* maintains an intracellular pH near neutrality under optimal conditions.[12] When host-cell-free *C. burnetii* are incubated at a pH level shown to activate metabolism without a metabolizable substrate, the intracellular pH is 5.88 whereas in the presence of glutamate, the intracellular pH rises to 6.95.[12] Intracellular pH changes of this magnitude or less have been shown to trigger such diverse metabolic events as sperm motility,[67] slime mold differentiation,[67] and *Bacillus* spore germination.[68] It is conceivable that nutrient depletion within the phagolysosome may result in an inability by *C. burnetii* to maintain intracellular pH homeostasis. This could potentially result in the activation of genes, such as *scvA*[38] or *hcbA*,[37] involved in morphogenesis of the SCV, the cell type adapted for extracellular survival.

An alternative proposal is that fluctuations in phagolysosomal pH may directly trigger pH-sensitive signal transduction systems present in the outer membrane of *C. burnetii*, leading to up- or downregulation of developmental genes. An adaptive sensory kinase has recently been identified in *C. burnetii*, but the environmental stimuli to which it responds are unknown.[69] The pH of vacuoles harboring *C. burnetii* within persistently infected cell lines remains relatively stable (approximately pH 4.8–5.0) over periods of many weeks.[7,11] However, the persistently infected cells employed in these studies were passaged three times weekly, thereby allowing normal cell cycle progression and division. This effectively reseeds the cell culture with uninfected daughter cells, which in turn become infected.[70] Whether alkalization of phagolysosomal pH occurs in long-term, heavily infected, nondividing cells has yet to be determined.

To test the hypothesis that changes in phagolysosomal pH directly effect *Coxiella* development, Vero cells infected for 48 hr were subsequently treated for 24 hr with bafilomycin A_1. Bafilomycin A_1 is a macrolide antibiotic that specifically inhibits the vacuolar-type (H+) ATPase (V-ATPase), a proton pump responsible for acidifying coated vesicles, endosomes, lysosomes, and the Golgi apparatus.[71] The pH of normally acidic vesicles rises to approximately 6.3 within 1 hr of treatment.[71] The V-ATPase has recently been localized by immunofluorescence to the *C. burnetii* vacuolar membrane.[8] After bafilomycin A_1 treatment, infected cells were viewed by TEM and *Coxiella* cell morphologies compared with the untreated control. There was no obvious difference in the ratio of SCV to LCV between treated and untreated cells. Moreover, increased production of SLP was not observed in treated cells. (R.A. Heinzen and T. Hackstadt, unpublished data, 1996) Although preliminary in nature, this study suggests that phagolysosomal pH is not an effector of *C. burnetii* development.

5.3. Temperature

Sudden temperature elevation induces a remarkable physiological response in both eukaryotic and prokaryotic cells termed *heat shock*. In this phenomenon cells undergo a massive redirection in gene expression, thus altering their pattern of protein synthesis and growth.[72] Proteins produced in response to thermal stress are termed *heat-shock proteins* (HSP) and presumably function in protecting the cell from the adverse environment.[72]

In several organisms, developmental control is superimposed on the transient activation of heat-shock genes, indicating possible dual roles for HSP in cellular differentiation and protection from adverse environmental conditions. For example, in *Xenopus*, Hsp70 is constitutively expressed without heat shock during oogenesis, but Hsp30 is not heat inducible during embryogenesis.[73] *Myxococcus xanthus* vegetative and fruiting cells exhibit different HSP profiles during thermal stress.[74] During early stages of sporulation in *Bacillus subtilis*, levels of some HSP increase without heat stress.[75]

Aerosol or tick transmission of *C. burnetii* to a mammalian host would present a heat-shock situation. This response has been observed *in vitro* during metabolic activation.[76] *C. burnetii* that are mechanically released from host cells display increased synthesis of at least seven proteins when shifted from 21 to 42°C during axenic metabolic activation.[76] Because SCV are considered the extracellular survival form of the bacterium, genes that are preferentially transcribed on uptake may include heat-shock genes that are activated during thermal stress and whose protein products are involved in morphogenesis, intracellular adaptation, or both.

6. CONCLUSIONS AND FUTURE PROSPECTS

Cellular differentiation in prokaryotes is an adaptive response that involves alterations in gene expression.[46–48] Morphological differentiation to a resistant, resting form is an obvious advantage for a pathogen like *C. burnetii* where the primary mode of disease transmission is by inhalation of contaminated aerosols from natural environments.[2] In addition, the antigenic variation displayed by SCV and LCV may partially explain the agent's ability to avoid the host's immune response and cause chronic infection.[16] Immunodominant antigens produced by LCV may be absent in SCV, and vice versa. This phenomenon has been demonstrated for the LCV-specific 29.5-kDa P1 protein.[16,30]

Many fundamental questions concerning the developmental biology of *C. burnetii* remain unanswered. Most importantly, is the SLP a spore that is shed by dying LCV and the infectious progenitor of the SCV? Or do SCV simply arise by differentiation and condensation of LCV in a scenario similar to chlamydial RB-to-EB transition? There is undoubtedly morphological differentiation from SCV to LCV. However, the temporal events of this morphogenesis are completely unknown as are the environmental conditions within the phagolysosomal milieu that drive development.

Studies investigating morphological differentiation and programmed gene expression in *C. burnetii* will be aided by transiently synchronizing the cell cycle. This can be accomplished by infecting with gradient-purified cell forms, or possibly by adding inhibitors to the culture medium that will drive a heterogeneous cell population toward one cell type. Cloning genes preferentially transcribed by SCV and LCV, and analyzing their products for function, may provide a first step in defining elements involved in morphogenesis. Such elements may include alternative sigma factors that function as transcriptional regulators of developmentally regulated genes in other organisms.[46,47] Cloning strategies can be multifaceted. Degenerate oligonucleotide probes corresponding to the N-terminal amino acid sequence of cell type-specific proteins can be utilized to retrieve the encoding genes from a genomic library. This approach was successfully employed in the cloning of *hcbA*[37] and *scvA*.[38] Polyclonal and monoclonal antibodies raised against SCV or LCV might similarly recognize unique cell type antigens and could be used as cloning reagents. In principle, purified SCV and LCV that are metabolically activated *in vitro*, should produce unique mRNA species. cDNA probes could then be generated from mRNA pools and used to retrieve developmentally regulated genes from a genomic library by utilizing a number of traditional screening methods.[77] This general approach was successfully employed in the cloning of a gene expressed early during *Chlamydia psittaci* EB-to-RB transition.[78] Obtaining adequate message from purified SCV may be problematic as the cell type displays diminished metabolic activity *in vitro*.[28] It may be possible to circumvent this obstacle by utilizing RNA extracted from host cells recently infected with purified SCV or LCV. Sophisticated techniques that couple the polymerase chain reaction (PCR) with subtraction methods, such as representational difference PCR,[79] could then be used to generate gene probes for differentially expressed SCV or LCV DNA sequences.

Considerable progress has been made in understanding temporal events in *Bacillus* development by analyzing mutants with specific developmental blocks.[46] The current lack of genetic systems of *Coxiella* precludes this type of investigation. A major advancement in this regard is a recent report by Suhan *et al.*[80] where they describe the introduction and expression of foreign DNA in *C. burnetii*. Employing a shuttle vector containing *E. coli* and *C. burnetii* replicons, they recently transformed *C. burnetii* to ampicillin resistance by electroporation. This exciting development will hopefully lead to workable systems for gene inactivation and subsequent mutant isolation, and provide a powerful tool in elucidating critical genetic events in *C. burnetii* morphogenesis.

ACKNOWLEDGMENTS. The author is indebted to T. Hackstadt for his contributions to this work. Acknowledgments also go to S. F. Hayes for electron microscopy, B. Evans and G. Hettrick for graphics, M. Peacock for propagation and purification of *C. burnetii*, and T. Hackstadt, D. Rockey, and S. Robertson for critical review of the manuscript.

REFERENCES

1. Baca, O. G., and Paretsky, D., 1983, Q-fever and *Coxiella burnetii:* A model for host–parasite interactions, *Microbiol. Rev.* **47:**127–149.
2. Babudieri, B., 1959, Q fever: A zoonosis, *Adv. Vet. Sci.* **5:**81–182.
3. Brouqui, P., Dupont, H. T., Drancourt, M., Berland, Y., Etienne, J., Leport, C., Goldstein, F., Massip, P., Micoud, M., Bertrand, A., and Raoult, D., 1993, Chronic Q fever, *Arch. Intern. Med.* **153:**642–648.
4. Relman, D. A., Lepp, P. W., Sadler, K. N., and Schmidt, T. M., 1992, Phylogenetic relationships among the agent of bacillary angiomatosis, *Bartonella bacilliformis,* and other alpha-proteobacteria, *Mol. Microbiol.* **6:**1801–1807.
5. Burton, P. R., Kordová, N., and Paretsky, D., 1971, Electron microscopic studies of the rickettsia *Coxiella burneti:* Entry, lysosomal response, and fate of rickettsial DNA in L-cells, *Can. J. Microbiol.* **17:**143–150.
6. Burton, P. R., Stueckemann, J., Welsh, R. M., and Paretsky, D., 1978, Some ultrastructural effects of persistent infections by rickettsia *Coxiella burnetii* in mouse L cells and green monkey kidney (Vero) cells, *Infect. Immun.* **21:**556–566.
7. Akporiaye, E. T., Rowatt, J. D., Aragon, A. A., and Baca, O. G., 1983, Lysosomal response of a murine macrophage-like cell line persistently infected with *Coxiella burnetii, Infect. Immun.* **40:**1155–1162.
8. Heinzen, R. A., Scidmore, M. A., Rockey, D. D., and Hackstadt, T., 1996, Differential interaction with endocytic and exocytic pathways distinguish the parasitophorous vacuoles of *Coxiella burnetii* and *Chlamydia trachomatis, Infect. Immun.* **64:**796–809.
9. Williams, J. C., 1991, Infectivity, virulence, and pathogenicity of *Coxiella burnetii* for various hosts, in: *Q Fever: The Biology of Coxiella burnetii* (J. C. Williams and H. A. Thompson, eds.), CRC Press, Boca Raton, pp. 21–71.
10. Baca, O. G., Klassen, D. A., and Aragon, A. S., 1993, Entry of *Coxiella burnetii* into host cells, *Acta Virol.* **37:**143–155.
11. Maurin, M., Benoliel, A. M., Bongrand, P., and Raoult, D., 1992, Phagolysosomes of *Coxiella burnetii*-infected cell lines maintain an acidic pH during persistent infection, *Infect. Immun.* **60:**5013–5016.
12. Hackstadt, T., and Williams, J. C., 1994, Metabolic adaptations of *Coxiella burnetii* to intra-phagolysosomal growth, in: *Microbiology 1984* (L. Lieve and D. Schlessinger, eds.), American Society for Microbiology, Washington, DC, pp. 266–268.
13. Reiner, N. E., 1994, Altered cell signaling and mononuclear phagocyte deactivation during intracellular infection, *Immun. Today* **15:**374–381.
14. Baca, O. G., Li, Y., and Kumar, H., 1994, Survival of the Q fever agent *Coxiella burnetii* in the phagolysosome, *Trends Microbiol.* **2:**476–480.
15. McCaul, T. F., and Williams, J. C., 1981, Developmental cycle of *Coxiella burnetii:* Structure and morphogenesis of vegetative and sporogenic differentiations, *J. Bacteriol.* **147:**1063–1076.
16. McCaul, T. F., 1991, The developmental cycle of *Coxiella burnetii,* in: *Q Fever: The Biology of Coxiella burnetii* (J. C. Williams and H. A. Thompson, eds.), CRC Press, Boca Raton, pp. 223–258.
17. Davis, G. E., and Cox, H. R., 1938, A filter-passing infectious agent isolated from ticks, *Public Health Rep.* **53:**2259–2267.
18. Kordová, N., 1959, Filterable particles of *Coxiella burneti, Acta Virol.* **3:**25–36.

19. Kordová, N., 1960, Study of antigenicity and immunogenicity of filterable particles of *Coxiella burneti*, *Acta Virol.* **4:**56–62.
20. Kordová, N., and Kovácová, E., 1968, Appearance of antigens in tissue culture cells inoculated with filterable particles of *Coxiella burneti* as revealed by fluorescent antibodies. *Acta Virol.* **12:**460–463.
21. Stocker, M. G. P., Smith, K. M., and Fiset, P., 1956, Internal structure of *Rickettsia burnetii* as shown by electron microscopy of thin sections, *J. Gen. Microbiol.* **15:**632–635.
22. Rosenberg, M., and Kordová, N., 1960, Study of intracellular forms of *Coxiella burneti* in the electron microscope, *Acta Virol.* **4:**52–55.
23. Anacker, R. L., Fukushi, K., Pickens, E. G., and Lackman, D. B., 1964, Electron microscopic observations of the development of *Coxiella burnetii* in the chick yolk sac, *J. Bacteriol.* **88:**1130–1138.
24. Nermut, M. V., Schramek, S., and Brezina, R., 1968, Electron microscopy of *Coxiella burneti* phase I and II, *Acta Virol.* **12:**446–452.
25. Wiebe, M. E., Burton, P. R., and Shankel, D. M., 1972, Isolation and characterization of two cell types of *Coxiella burnetii* phase I, *J. Bacteriol.* **110:**368–377.
26. Canonico, P. G., Van Zwieten, M. J., and Christmas, W. A., 1972, Purification of large quantities of *Coxiella burnetii* rickettsia by density gradient zonal centrifugation, *Appl. Microbiol.* **23:**1015–1022.
27. Wachter, R. F., Briggs, G. P., Gangemi, J. D., and Pedersen, Jr., C. E., 1975, Changes in buoyant density relationships of two cell types of *Coxiella burnetii* phase I, *Infect. Immun.* **12:**433–436.
28. McCaul, T. F., Hackstadt, T., and Williams, J. C., 1981, Ultrastructural and biological aspects of *Coxiella burnetii* under physical disruptions, in: *Rickettsiae and Rickettsial Diseases* (W. Burgdorfer and R. L. Anacker, eds.), Academic Press, New York, pp. 267–280.
29. McCaul, T. F., and Williams, J. C., 1990, Localization of DNA in *Coxiella burnetii* by post-embedding immunoelectron microscopy, *Ann. N.Y. Acad. Sci.* **590:**136–147.
30. McCaul, T. F., Banerjee-Bhatnager, N., and Williams, J. C., 1991, Antigenic differences between *Coxiella burnetii* cells revealed by postembedding immunoelectron microscopy and immunoblotting, *Infect. Immun.* **59:**3243–3253.
31. McCaul, T. F., Williams, J. C., and Thompson, H. A., 1991, Electron microscopy of *Coxiella burnetii* in tissue culture. Induction of cell types as products of developmental cycle, *Acta Virol.* **35:**545–556.
32. McCaul, T. F., Dare, A. J., Gannon, J. P., and Galbraith, A. J., 1994, In vivo endogenous spore formation by *Coxiella burnetii* in Q fever endocarditis, *J. Clin. Pathol.* **47:**978–981.
33. Amano, K., Williams, J. C., McCaul, T. F., and Peacock, M. G., 1984, Biochemical and immunological properties of *Coxiella burnetii* cell wall and peptidoglycan–protein complex fractions, *J. Bacteriol.* **160:**982–988.
34. Kishimoto, R. A., Veltri, B. J., Canonico, P. G., Shirey, F. G., and Walker, J. S., 1976, Electron microscopic study on the interaction between normal guinea pig peritoneal macrophages and *Coxiella burnetii*, *Infect. Immun.* **14:**1087–1096.
35. Avakyan, A. A., Popov, V. L., Chebanov, S. M., Shatkin, A. A., Siderov, V. E., and Kudelina, R. I., 1983, Comparison of the ultrastructure of small dense forms of *Chlamydiae* and *Coxiella burnetii*, *Acta Virol.* **27:**168–172.
36. Schaal, F., Krauss, H., Jekov, N., and Rantamäki, L., 1987, Electron microscopic observations on the morphogenesis of "spore-like particles" of *Coxiella burnetii* in cell cultures, *Acta Medit. Patol. Inf. Trop.* **6:**329–338.
37. Heinzen, R. A., and Hackstadt, T., 1996, A developmental stage-specific histone H1 homolog of *Coxiella burnetii*, *J. Bacteriol.* **178:**5049–5052.
38. Heinzen, R. A., Howe, D., Mallavia, L. P., Rockey, D. D., and Hackstadt, T., 1996, Developmentally regulated synthesis of an unusually small, basic peptide by *Coxiella burnetii*, *Mol. Microbiol.* **22:**9–19.

39. Zuerner, R. L., and Thompson, H. A., 1983, Protein synthesis by intact *Coxiella burnetii* cells, *J. Bacteriol.* **156:**186–191.

40. Thompson, H. A., Zuerner, R. L., and Redd, T., 1984, Protein synthesis in *Coxiella burnetii*, in: *Microbiology 1984* (L. Lieve and D. Schlessinger, eds.), American Society for Microbiology, Washington, DC, pp. 288–292.

41. Redd, T., and Thompson, H. A., 1995, Secretion of proteins by *Coxiella burnetii*, *Microbiology* **141:**363–369.

42. Small, P. L. C., Ramakrishnan, L., and Falkow, S., 1994, Remodeling schemes of intracellular pathogens, *Science* **263:**637–639.

43. Schachter, J., 1988, The intracellular life of *Chlamydia*, *Curr. Top. Microbiol. Immun.* **138:**109–139.

44. Hackstadt, T., Baehr, W., and Ying, Y., 1991, *Chlamydia trachomatis* developmentally regulated protein is homologous to eukaryotic histone H1, *Proc. Natl. Acad. Sci. USA* **88:**3937–3941.

45. Brickman, T. J., Barry, C. E., III, and Hackstadt, T., 1993, Molecular cloning and expression of *hctB* encoding a strain-variant chlamydial histone-like protein with DNA-binding activity, *J. Bacteriol.* **175:**4274–4281.

46. Errington, J., 1993, *Bacillus subtilis* sporulation: Regulation of gene expression and control of morphogenesis, *Microbiol. Rev.* **57:**1–33.

47. Chater, K. F., and Hopwood, D. A., 1993, *Streptomyces*, in: *Bacillus subtilis and Other Gram-Positive Bacteria* (A. L. Sonenshein, J. A. Hoch, and R. Losick, eds.), American Society for Microbiology, Washington, DC, pp. 83–99.

48. Kaiser, D., 1986, Regulation of multicellular development in myxobacteria, in: *Microbial Development* (R. Losick and L. Shapiro, eds.), Cold Spring Harbor Laboratory, Cold Spring Harbor, pp. 197–216.

49. Drlica, K., and Rouviere-Yaniv, J., 1987, Histonelike proteins of bacteria, *Microbiol. Rev.* **51:**301–319.

50. Musgrave, D. R., Sandman, K. M., and Reeve, J. N., 1991, DNA binding by the archaeal histone HMf results in positive supercoiling, *Proc. Natl. Acad. Sci. USA* **88:**10397–10401.

51. Setlow, P., 1988, Small, acid-soluble spore proteins of *Bacillus* species: Structure, synthesis, genetics, function and degradation, *Annu. Rev. Microbiol.* **42:**319–338.

52. Nicholson, W. L., Setlow, B., and Setlow, P., 1990, Binding of DNA in vitro by a small, acid-soluble protein from *Bacillus subtilis* and the effect of this binding on DNA topology, *J. Bacteriol.* **172:**6900–6906.

53. Setlow, B., and Setlow, P., 1995, Small, acid-soluble proteins bound to DNA protect *Bacillus subtilis* spores from killing by dry heat, *Appl. Environ. Microbiol.* **61:**2787–2790.

54. Williams, J. C., 1991, Antigens, virulence factors, and biological response modifiers of *Coxiella burnetii*: Strategies for vaccine development, in: *Q Fever: The Biology of Coxiella burnetii* (J. C. Williams and H. A. Thompson, eds.), CRC Press, pp. 175–222.

55. Sugawara, E., and Nikaido, H., 1992, Pore-forming activity of OmpA protein in *Escherichia coli*, *J. Biol. Chem.* **267:**2507–2511.

56. Williams, J. C., Hoover, T. A., Waag, D. M., Banerjee-Bhatnagar, N., Bolt, C. R., and Scott, G. H., 1990, Antigenic structure of *Coxiella burnetii*: A comparison of lipopolysaccharide and protein antigens as vaccines against Q fever, *Ann. N.Y. Acad. Sci.* **590:**370–380.

57. Campbell, S., Richmond, S. J., Yates, P. S., and Story, C. C., 1994, Lipopolysaccharide in cells infected by *Chlamydia trachomatis*, *Microbiology* **140:**1995–2002.

58. Aronson, A. T., and Fitz-James, P., 1976, Structure and morphogenesis of the bacterial spore coat, *Bacteriol. Rev.* **40:**360–402.

59. Oswald, W., and Thiele, D., 1993, A sporulation gene in *Coxiella burnetii? J. Vet. Med.* **40:**366–370.

60. Foulger, D., and Errington, J., 1989, The role of the sporulation gene *spoIIIE* in the regulation of prespore-specific gene expression in *Bacillus subtilis*, *Mol. Microbiol.* **3:**1247–1255.

61. VanDemark, P. J., and Batzing, B. L., 1987, *The Microbes: An Introduction to Their nature and Importance*, Benjamin/Cummings, Menlo Park, CA.
62. Amano, K., Hatakeyama, H., Sasaki, Y., Ito, R., Tamura, A., and Suto, T., 1991, Electron microscopic studies on the in vitro proliferation of spotted fever group rickettsia isolated in Japan, *Microbiol. Immunol.* **35:**623–629.
63. Kölbel, H. K., 1984, Electron microscopy, in: *The Mycobacteria, A Sourcebook*, Part A (G. P. Kubica and L. G. Wayne, eds.), Dekker, New York, pp. 249–300.
64. Socolofsky, M. D., and Wyss, O., 1961, Cysts of Azotobacter, *J. Bacteriol.* **81:**946–954.
65. Paul, T. R., and Beveridge, T. J., 1992, Reevaluation of envelope profiles and cytoplasmic ultrastructure of mycobacteria processed by conventional embedding and freeze-substitution protocols, *J. Bacteriol.* **174:**6508–6517.
66. Olson, E. R., 1993, Influence of pH on bacterial gene expression, *Mol. Microbiol.* **8:**5–14.
67. Busa, W. B., and Nuccitelli, R., 1984, Metabolic regulation via intracellular pH, *Am. J. Physiol.* **246:**R409–R438.
68. Setlow, B., and Setlow, P., 1980, Measurements of the pH within dormant and germinated bacterial spores, *Proc. Natl. Acad. Sci. USA* **77:**2474–2476.
69. Mo, Y., and Mallavia, L. P., 1994, A *Coxiella burnetii* gene encodes a sensor-like protein, *Gene* **151:**185–190.
70. Roman, M. J., Coriz, P. D., and Baca, O. G., 1986, A proposed model to explain persistent infection of host cells with *Coxiella burnetii*, *J. Gen. Microbiol.* **132:**1415–1422.
71. Yoshimori, T., Yamamoto, A., Moriyama, Y., Futai, M., and Tashiro, Y., 1991, Bafilomycin A₁, a specific inhibitor of vacuolar-type H+-ATPase, inhibits acidification and protein degradation in lysosomes of cultured cells, *J. Biol. Chem.* **266:**17707–17712.
72. Neidhardt, F. C., and VanBogelen, R. A., 1987, Heat shock response, in: *Escherichia coli and Salmonella typhimurium. Cellular and Molecular Biology* (F. C. Neidhart, ed.), American Society of Microbiology, Washington, DC, pp. 1334–1345.
73. Bienz, M., 1984, Developmental control of the heat shock response in *Xenopus*, *Proc. Natl. Acad. Sci. USA* **81:**3138–3142.
74. Nelson, D. R., and Killeen, K. P., 1986, Heat shock proteins of vegetative and fruiting *Myxococcus xanthus* cells, *J. Bacteriol.* **168:**1100–1106.
75. Todd, J. A., Hubbard, T. J. P., Travers, A. A., and Ellar, D. J., 1985, heat-shock proteins during growth and sporulation of *Bacillus subtilis*, *FEBS Lett.* **188:**209–214.
76. Thompson, H. A., Bolt, C. R., Hoover, T., and Williams, J. C., 1990, Induction of heat-shock in *Coxiella burnetii*, *Ann. N.Y. Acad. Sci.* **590:**127–135.
77. Chisholm, R. L., 1987, Isolation of developmentally regulated genes, *Methods Cell Biol.* **28:**461–470.
78. Wichlan, D. G., and Hatch, T. P., 1993, Identification of an early-stage gene of *Chlamydia psittaci* 6BC, *J. Bacteriol.* **175:**2936–2942.
79. Hubank, M., and Schatz, D. G., 1994, Identifying differences in mRNA expression by representational difference analysis of cDNA, *Nucleic Acids Res.* **22:**5640–5648.
80. Suhan, M. L., Shu-Yin, C., and Thompson, H. A., 1996, Transformation of *Coxiella burnetii* to ampicillin resistance, *J. Bacteriol.* **178:**2701–2708.

8

The Identification of Virulence Factors of *Coxiella burnetti*

OSWALD G. BACA and LOUIS P. MALLAVIA

1. INTRODUCTION

Coxiella burnetii is an obligate intracellular bacterium that causes Q fever in humans and is acquired through the inhalation of contaminated dust particles or by contact with infected domestic animals such as cattle and sheep.[1] Typical of Q fever are flulike symptoms accompanied by an intense preorbital headache and fever; recovery usually occurs within 1–2 weeks.[2] Acute disease may progress to lifethreatening chronic Q fever which frequently manifests as endocarditis involving the aortic and/or mitral valves.[3]

After attachment to the surface of a variety of target cells (e.g., fibroblasts, endothelial cells, macrophages), *C. burnetii* passively enters via phagocytosis.[4] Once inside, the organism localizes and proliferates within the confines of the acidic phagolysosome (Fig. 1).[5–7] Heavily infected cells typically exhibit only one infected vacuole, which suggests that parasite-containing vacuoles may fuse.[8,9] Such heavily infected cells also can divide giving rise to infected and uninfected daughter cells.[8] The acid environment is required by the organism for transporting glucose and other biochemicals across its membrane.[10] *In vitro* studies with J774 macrophage cells have revealed that antibody (IgG) to *C. burnetii* does not

This review is dedicated to David Paretsky, scientist, teacher, and friend.

OSWALD G. BACA • Department of Biology, The University of New Mexico, Albuquerque, New Mexico 87131.   LOUIS P. MALLAVIA • Department of Microbiology, Washington State University, Pullman, Washington 99164.

Rickettsial Infection and Immunity, edited by Anderson *et al.* Plenum Press, New York, 1997.

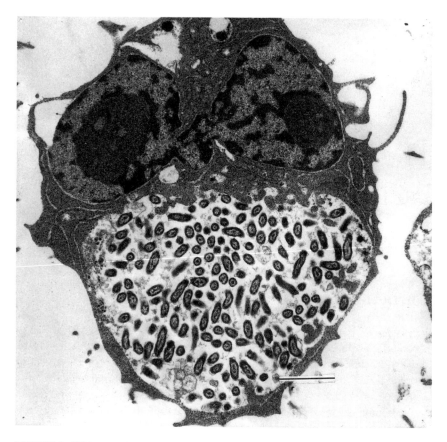

FIGURE 1. J774 macrophage with *Coxiella burnetti*, proliferating within the phagolysosome. Bar = 2 μm. (From Ref. 11, with permission.)

prevent infection and that it accelerates entry of the agent.[11] Exposure of *C. burnetii* to a synthetic antibacterial peptide (CAP37$_{20-44}$) based on the active region of CAP37, a cationic protein present in the lysosomal granules of human neutrophils, has no adverse effect on the parasite; in fact, the peptide enhances infection of host cells.[12] Cell-mediated immunity is probably ultimately responsible for controlling the disease with humoral immunity playing an ancillary role (reviewed in Ref. 1).

    The properties of *C. burnetii* that allow it to escape destruction within the harsh environment of the phagolysosome have not been conclusively identified; however, the parasite has several enzymes—superoxide dismutase (SOD), catalase, and acid phosphatase—that may allow it to survive within the phagolysosome. The microbe's surface lipopolysaccharide (LPS) is a virulence factor that may, directly or indirectly, contribute to its survival. Strain designations have been proposed for *C. burnetii* based on genomic restriction fragment polymorphisms and plasmid content.[13–15] The hypothesis that some of these strains may be associated with distinct disease states (acute or chronic) suggests that the plasmids encode unique virulence factors.[13,15] Indeed, plasmid-encoded proteins expressed on the parasite's surface have been described which may be associated with virulence.

In this review we examine these enzymes and surface components and other potential virulence factors that may promote the parasite's capacity to establish itself within the host cell. Also discussed in this review is the hypothesis that different strains of *C. burnetii*—differentiated on the basis of plasmid content and type—may be responsible for either acute or acute/chronic disease states. Recent data that do not support this hypothesis are also examined. The application of molecular genetic approaches to the search for virulence factors is crucial for elucidating the virulence factors of *C. burnetii;* indeed, these approaches are clarifying the nature of previously identified potential factors (e.g., SOD and LPS) and are revealing other potentially significant virulence determinants such as macrophage infectivity potentiator and an outer membrane protein, Com1, which may mediate attachment/entry into host cells.

2. SOD AND CATALASE

Professional phagocytes generate a number of toxic oxygen metabolites, including superoxide anion, during the enhanced metabolism (metabolic burst) that accompanies the uptake of microorganisms. Several pathogenic bacteria avoid the killing effect of superoxide by destroying it with the aid of SOD and catalase.[16] SOD, through a dismutation reaction, converts the toxic superoxide anion to hydrogen peroxide; catalase, in turn, catalyzes the production of H_2O and oxygen from H_2O_2. *C. burnetii* possesses both enzymes.[17] The SOD-encoding gene of *Coxiella* has been cloned and functionally expressed in an *Escherichia coli* SOD-deficient mutant.[18] Inactivation of the cloned enzyme by hydrogen peroxide and comparison of its deduced amino acid sequence with known SODs indicate that it is an iron-containing enzyme.

C. burnetii generates significant levels of superoxide anion,[17] and it is likely that its SOD and catalase eliminate this potentially toxic molecule. The pH at which the enzymes function optimally—pH 7.8 (SOD) and 7.0 (catalase)[17]— indicates that they are cytoplasmic enzymes (the intracellular pH of *C. burnetii* is approximately 7.0) and that they eliminate internally generated superoxide. It is highly unlikely that the enzymes are exported into the external acidic environment of the phagolysosome as the activities of the two are greatly diminished at pH 4.5.[17]

Whether or not SOD of *C. burnetii* protects it from exogenous superoxide (i.e., produced by the host cell) is unknown. The dogma[19] that external superoxide does not cross membranes has been questioned as a result of recent experiments with fungi.[20,21] Increased spontaneous mutation rates were observed in SOD mutants of *Saccharomyces cerevisiae* and *Neurospora crassa* grown under aerobic conditions and were similar to the mutation rates of the wild-type cells cultivated under anaerobic conditions; these results clearly imply that superoxide generated during respiration penetrates through the mitochondrial and nuclear membranes and subsequently damages nuclear DNA.

Because neutral compounds such as hydrogen peroxide are capable of traversing membranes,[22] it is likely that catalase protects *C. burnetii* from exogenous hydrogen peroxide that penetrates the organism's envelope.

3. ACID PHOSPHATASE (ACP)

Successful parasitization of phagocytes by the Q fever agent may be related, in part, to the parasite's suppression of the phagocyte's capacity to generate adequate concentrations of microbicidal oxygen metabolites during and after parasite entry. Several years ago it was noted that macrophage cells (J774 cell line) infected with *C. burnetii* exhibited a diminished level of oxidative metabolism and reduced capacity to generate superoxide anion.[11] Moreover, although oxidative metabolism occurred in J774 cells during the phagocytosis of antibody-treated *C. burnetii*, it could not be detected in cells phagocytosing unopsonized rickettsiae. More recently, it was found that phagocytosis of either opsonized or unopsonized *C. burnetii* failed to trigger a significant production of the superoxide anion by human neutrophils.[23] Similar results with neutrophils were also reported by Ferencik *et al.*[24]

Both *Legionella micdadei* and *Leishmania* promastigotes have ACPs that inhibit the respiratory (oxidative) burst of neutrophils, apparently by reducing the amount of the second messengers, inositol 1,4,5-triphosphate (IP_3) and *sn*-1,2-diacylglycerol, produced following receptor-mediated stimulation.[25,26] These facultative intracellular parasitic agents proliferate in phagocytes; *Leishmania* grows in phagolysosomes. Recently it was discovered that *C. burnetii* also possess significant ACP activity which is probably responsible for inhibiting the metabolic burst in human neutrophils.[27] Soluble fractions derived from sonicated *C. burnetii* contain considerable ACP activity, which blocks superoxide-anion production by human neutrophils stimulated with the peptide formyl-Met-Leu-Phe; progressive purification (centrifugation, column chromatography, ultrafiltration, and high-perfor-

mance liquid chromatography) of the ACP to near homogeneity (100-fold purification) resulted in enhanced neutrophil inhibition. Its molecular mass (determined by SDS-PAGE) is approximately 91 kDa; its pH optimum is approximately 5.0. Interestingly, the ACP is apparently glycosylated as treatment with
N-glycohydrolase reduced its molecular mass by approximately 3 kDa.[100] Phosphatase activity has been detected in all isolates of *C. burnetii* examined, including
those implicated in acute (Nine Mile) and chronic (S Q217, PRS Q177, K Q154)
disease. The level of activity is significantly higher than that reported in other
microorganisms, including *Leishmania* and *Legionella*.[27]

Cytochemical techniques coupled with electron microscopy localized the phosphatase activity to the periplasmic gap of *C. burnetii*. Ion-exchange chromatography
revealed that *C. burnetii* has a major species of the enzyme that is distinct from that of
the host cell. The phosphatase-inhibiting molybdate heteropolyanion complexes B'
$\{[C(NH_2)_3]_2[(CH_3)_2AsMo_4O_{15}H]\}$ and E_2 $[(NH_4)_6As_2Mo_{18}O_{62}\times H_2O]$ block
this inhibition. Treatment of persistently infected L929 cells with complex B'
significantly reduces the number of intracellular *C. burnetii* without affecting the
growth of the host cell.[27] These results suggest a possible strategy for treating
chronically infected patients, a group with a high mortality rate even with antibiotic
treatment or surgical intervention.

These data suggest that the parasite's ACP may be a major virulence determinant, allowing *C. burnetii* to avoid killing during uptake by phagocytes and,
later, in the phagolysosome. The data clearly show that the ACP is parasite-
encoded. Where the ACP is encoded—plasmid or chromosome—has not been
determined. Substrate specificity studies suggest that the ACP may target phosphotyrosine-containing proteins[100]—such proteins have been implicated in the
signal transduction process associated with phagocytosis (see below). The *Coxiella*
ACP dephosphorylates Raytide, the phosphotyrosine-containing peptide used in
determining substrate specificity of phosphatases.

Protein tyrosine phosphatase activity has also been reported in *Yersinia* and it
has been shown that this secreted, plasmid-encoded enzyme is essential for pathogenesis. The phosphatase apparently inhibits the Fc-receptor-mediated signal
transduction process in macrophages which culminates in the oxidative burst.[28]

In *Salmonella*, expression of ACP-encoding genes is regulated by a pair of
genes, *phoQ* and *phoP*, that are homologous to a family of two-component regulatory systems.[29-31] Moreover, in *S. typhimurium*, *phoQ* and *phoP* have been implicated as virulence determinants enhancing survival in macrophages.[30,32] Two-
component regulatory systems have been identified in a large number of bacterial
species and shown to regulate a variety of cellular processes.[31] These systems
consist of at least two components: One is a histidine protein kinase family, sensor
component; its N-terminal domain extending into the periplasmic gap interacting
with stimulatory ligands which act to control the kinase (or phosphatase) activities

of the cytoplasmic C-terminal domain. The other is a response regulator family component; phosphorylation of its N-terminal domain by the sensor, stimulates its ability to bind DNA and is required for activation of transcription.[31] As a first step toward identifying genes regulating ACP production in *C. burnetii*, a gene encoding a sensor protein with similarity to a PhoQ analogue has been isolated by PCR amplification using degenerate primers to conserved residues of members of the histidine protein kinase family. Sequencing revealed an ORF that encodes a protein of 425 amino acids (48 kDa). The ORF shares significant homology with sensor proteins and has been designated QrsA (for *Q* fever agent *r*egulatory *s*ensorlike protein).[33] QrsA shares overall amino acid similarity with CpxA (50.5%), PhoR (48.6%), and PhoQ (47.2%). These sensor proteins appear to be involved in sensing PO_4 levels, and in the case of PhoQ, inducing acid phosphatase (ACP). *C. burnetii qrsA* gene sequences are present in all genomic groups tested. Preliminary results have demonstrated the presence of an ORF encoding the putative response regulator (DNA binding protein) component of this system.

4. MACROPHAGE INFECTIVITY POTENTIATOR (MIP)

In searching for virulence factors of *Legionella pneumophila*, Cianciotto *et al.*[34] identified a gene encoding a 24-kDa surface protein involved in macrophage infection which they designated *lpmip* (*L. pneumophila* macrophage infectivity potentiator). Since then, *mip*-like genes have been cloned from other intracellular bacteria including *Chlamydia trachomatis,* and the gene products shown to possess amino acid sequence similarity to LpMip.[35-38] Genetic data indicate that Mip plays an important role as a virulence factor in *L. pneumophila*, as mutations in the *mip* gene cause a reduction in intracellular infection.[36,39,40] Previous studies revealed that *C. burnetii* DNA contained sequences with homology to the *LpMip* gene.[41] Recently, using monospecific polyclonal antibodies to LpMip, a *C. burnetii* gene encoding a Mip-like analogue has been cloned and sequenced.[42] The *C. burnetii* gene has been named *cbmip* (*C. burnetii macrophage infectivity potentiator*). *The CbMip protein has overall sequence similarity to LpMip (46%),* a *Chlamydia trachomatis* Mip (CtMip) (30%), and the C-terminus (PPIase region) has over 35% identify to FK506 binding proteins (FKBP) found in prokaryotic and eukaryotic cells. Like FKBPs, CbMip has peptidyl-prolyl isomerase (PPIase) activity that is inhibited by rapamycin.[42]

Preliminary data indicate that CbMip is surface accessible to proteinase K digestion of intact cells. Of considerable interest is the possibility that CbMip may be an exported protein that is excreted into the host phagolysosomal vacuole and perhaps into the host cytoplasm.[43] CbMip appears to be an immunoreactive protein because mice infected with *C. burnetii* develop antibodies to CbMip.

Although the role of bacterial Mip-like proteins is not known, LpMip has been shown to inhibit phosphokinase C (PKC) activity of neutrophils.[44] Further, rapamycin has been demonstrated to inhibit growth of *Chlamydia trachomatis* when added to the cells either prior to or at the time of infection.[45] FKBP–rapamycin complexes appear to play a role in modulating T cells, resulting in the downregulation of effects of IL-2 and IFNγ on T cells.[46-48] Further, the FKBP–FK506 drug complex appears to act by inhibition of the calcium regulator, calcineurin (calcium/calmodulin-dependent serine/threonine phosphatase).[48,49]

It will be interesting to determine if CbMip has any direct or indirect effects on T cells as preliminary data show that both IL-2 and IFNγ levels are reduced in the *C. burnetii*-infected mouse (J. Seshu, manuscript in preparation). FKBPs appear to play a role in modulating T cells, resulting in the downregulation of IL-2 and IFNγ synthesis.[48,50] CbMip could be very important in modulation of the host immune response leading to persistent infection and possibly to chronic disease.

5. COM1

Because the identification of virulence components by the isolation of mutants is difficult in *C. burnetii*, an approach that has been used to isolate surface-exposed putative virulence factors is to clone immunoreactive surface proteins. Using this approach a *C. burnetii* gene designated *com1* was isolated and expressed in *E. coli*.[51,52] The ORF for this gene encodes a predicted protein of 27.6 kDa. The protein seen in iodinated whole cells of *E. coli* carrying the *C. burnetii* gene or *C. burnetii* had a molecular weight of 25,700. This suggested the presence of a leader sequence. Leader sequences are usually associated with proteins destined for the outer membrane, the cytoplasmic membrane, or the periplasmic space.[53] The Com1 protein was shown to be an outer membrane protein. The *com1* gene is present in all isolates tested, thus it appears to encode a common antigen and may play a role in virulence such as attachment or entry. Computer data base analyses reveal that it has sequence homology to protein disulfide oxidoreductase. Protein disulfide oxidoreductases catalyze thiol-disulfide interchange reactions, promoting protein disulfide formation, reduction, or isomerization, depending on the redox potential and the polypeptide substrate.[36,54,55] Com1 may be released from the surface of *C. burnetii* in the phagolysosomal environment, affecting the surface proteins of *C. burnetii* or proteins in the phagolysosome. One potential target could be the defensin-like cationic polypeptides found in the phagolysosome of phagocytic cells, compounds that do not appear to affect *C. burnetii*.[12] Alternatively, Com1 may be needed to cross-link surface proteins of *C. burnetii* as it goes through the different stages of its developmental life cycle (see Heinzen, this volume).

6. OTHER SURFACE PROTEINS

With the exception of plasmid-specified proteins (see below), there are no obvious differences between strains, as seen by SDS-PAGE. A 29.5-kDA protein which can be surface iodinated and immunoprecipitated with immune sera has been purified by high-performance liquid chromatography from detergent extracts of *C. burnetii*.[56] This protein was closely associated with the cell wall and with peptidoglycan.[57] A protein of 27 kDa is the major labeled peptide in a membrane fraction of *C. burnetii* cells harvested from tissue culture media of infected cells. This protein was not found in organisms harvested from disrupted, infected tissue culture cells.[58] Finally, one group of researchers had reported using a highly purified, 27-kDa *C. burnetii* protein as a vaccine for guinea pigs, mice, and cattle.[59,60] The role, if any, of any of these proteins in the pathogenesis of *C. burnetii* is presently unknown.

7. LPS

LPS, which is typically associated with gram-negative bacteria, is also present in *C. burnetii*. The *C. burnetii* LPS has the polysaccharide complexity that is typically associated with LPSs of other bacteria; it also contains a lipid A moiety with a variety of fatty acids.[61–63] *C. burnetii* LPS varies antigenically and biochemically between strains.[64] Virulent organisms (designated phase I and found in nature) have a "smooth"-type LPS, whereas avirulent phase II (laboratory-derived through serial passage in embryonated eggs) rickettsiae have a truncated LPS typical of "rough" organisms.[65,66] The carbohydrate portion of phase II LPS lacks many of the sugars present in phase I LPS.[61,65,66] Compared with phase I *C. burnetii*, avirulent phase II organisms contain approximately 10% of the extractable LPS.[66]

These quantitative and qualitative differences may account for the differential uptake of phase I and phase II organisms by host cells: phase II *C. burnetii* are more readily phagocytosed than are phase I.[4] Qualitative differences in the LPS of *S. typhimurium* variants have been correlated with differential uptake by phagocytic cells[68]—variants with shorter LPS are more readily phagocytosed. The lower amount of LPS possessed by phase II organisms and the reduced carbohydrate content of the LPS probably enhance the rickettsial surface hydrophobicity, which, in turn, results in greater affinity for other hydrophobic structures such as host membranes. Whether or not the biochemical nature of the LPS determines if the organism can survive within the phagolysosome is unknown. However, both phase I and phase II *C. burnetii* proliferate within and establish a persistent infection of several murine macrophage cell lines (P388D1, J774, PU-5-IR),[69] suggesting that the nature of the LPS is not critical.

Vishwanath and Hackstadt[70] presented convincing data that phase I organisms resist complement-mediated serum killing, whereas the phase II variants are killed by complement. The smooth LPS variant (phase I) activates complement inefficiently and does not bind the opsonin C3b, whereas the rough LPS variant (phase II) activates complement via the alternative pathway. These results explain, in part, the avirulent nature of phase II organisms which appears to be correlated with LPS structure; such a correlation was previously demonstrated for gram-negative enteric bacteria.[71]

Another possible role proposed for phase I *Coxiella* LPS by Hackstadt is the masking of underlying antigens from the immune system, allowing the organism to escape immune surveillance during the early stages of infection.[72]

Whether or not the *Coxiella* LPS acts as an endotoxin is debatable because the amount of LPS required to induce toxicity in experimental animals is greater than for other gram-negative bacteria.[63,65,73] Guinea pigs exposed to LPS (~ 100 µg) derived from phase I *Coxiella* induced toxic responses typical of classic endotoxemia, including hyperthermia, weight loss, hepatomegaly with concomitant lipid infiltration, elevated cortisol levels in plasma and liver, and leukocytosis;[73] LPS from phase II *Coxiella* did not produce a toxic response.[67] Similar amounts of phase I LPS killed 50% of test chicken embryos which is approximately 3000 times the dose required with *Salmonella* LPS.[65] Phase I LPS elicits the dermal Shwartzman reaction (hemorrhaging, necrosis) typical of classic endotoxin, causes tumors to regress, and induces nonspecific immunity in mice to virulent yeast (reviewed in Ref. 1). Schramek *et al.*[74] showed that live or killed *C. burnetii* enhanced mouse lethality caused by injection of *Salmonella abortus-equi* LPS which is reminiscent of the effects of other macrophage stimulators, such as bacillus Calmette–Guérin. The same workers proposed the notion that through some unknown mechanism, *C. burnetii* may enhance the endotoxic activity of its own LPS.

8. ASSOCIATION BETWEEN PLASMIDS AND TYPE OF DISEASE

Although the development of chronic infections such as endocarditis has been considered a result of the compromised nature of the host and not a specific property of the pathogen,[75–77] more recent studies suggested that organisms causing endocarditis differed from those causing acute infections, with respect to LPS,[64,65,78] plasmid type,[15,79,80] and genomic restriction fragment length polymorphisms (RFLP) as determined by PAGE[13] and pulsed field gradient electrophoresis (PFGE).[81–83]

Studies on plasmid distribution led to the conclusion that different strains of
C. burnetii may be responsible for the acute and chronic manifestation of Q
fever.[15] Restriction-endonuclease mapping of *C. burnetii* plasmids from several
isolates suggested that plasmid type and disease state are associated. These observations resulted in a proposed classification[14] of *C. burnetii* into several major
categories or strains, including one associated with acute disease and two others
associated with chronic disease (i.e., primarily endocarditis). "Acute strain" organisms contain a 36-kb plasmid (QpH1), whereas one of the two strains associated with chronic disease contains a similar, but 3 kb larger, plasmid (QpRS).[15]
The third strain has a chromosomally integrated QpRS-type plasmid.[14] Similar
correlations between strains and disease manifestations have also been proposed
based on antigenic variation of *C. burnetii LPS*.[64] Antibiotic susceptibility testing of
a limited number of isolates has revealed that isolates associated with chronic Q
fever are more resistant to a variety of antibiotics.[84]

Because of the apparent connection between plasmid and disease type,
emphasis has been placed on the differences in plasmids. The examination of
these plasmids has led to the identification and characterization of a partition
region which ensures stable distribution of the plasmid to daughter cells.[85,86] In
addition, an open reading frame (ORF) with homology to those in chromosomal
replication regions has also been described.[87] Although QpRS and QpH1 plasmids have regions of considerable sequence similarity, they also contain unique
sequences.[14] The possibility exists that the plasmids encode virulence properties;
to date, however, none have been described, although plasmid-encoded proteins
of unknown function have been characterized. A gene termed *cbhE'* has been
cloned from the QpH1 plasmid and a 42-kDa polypeptide (CbhE) was expressed
in vitro and in *E. coli*.[88] The gene was only detected in strains associated with acute
disease. A surface-exposed protein (E') encoded within the *cbhE'* gene of the
QpRS plasmid has also been described by this group and is only associated with
the chronic disease-causing *C. burnetii* strain.[89]

Recent studies have cast doubt on the association of plasmid type and
disease manifestation[90-93] and the entire sequence of QpH1 is now known and
reveals no clues as to virulence.[94] Studies by a group in France have found no
correlation between the presence of the QpH1 plasmid (or the *cbhE'* gene) and
disease state.[90] *Coxiella* isolates derived from patients with either chronic or acute
disease were examined for the presence of the QpH1 plasmid and *cbhE'* gene.
Several of the isolates from chronically ill patients contained the QpH1-specific
cbhE' gene that Mallavia's group had reported as being unique to the QpH1
plasmid and present only in acute-disease-causing strains. Furthermore, of seven
isolates obtained from patients with acute Q fever, only two had the *cbhE'* gene.
Some of these more recent isolates along with the initial proposed strain designations are shown in Table I.

TABLE I
Proposed *Coxiella burnetii* Strain Designations

Genomic group	Representative isolates	Strain	Disease in humans	Plasmid type	Plasmid size
I	Nine Mile RSA493 Ohio 314 RSA338 Turkey RSA333	Hamilton	Acute	QpH1	36 kb
II	Henzerling M-44 RSA459	Vacca	Acute	QpH1	36 kb
III	Koka Idaho goat	Rasche	Acute	QpH1	36 kb
IV	Priscilla Q177 K Q154 F Q228	Biotzere	Chronic	QpRS	39 kb
V	L Q216 S Q217	Corazon	Chronic	Plasmidless	None[a]
VI	Dugway[b]	Dod	Unknown	QpDG	>42 kb
VII	Q321[c], Q1140, ME MAN, MAC	Iris	Acute, chronic	QpDV	33.5 kb
VIII Others?	CBQY[d] European isolates	?	Chronic	?	56 kb

[a]Contains sequences with homology to all four *Coxiella* plasmids.
[b]13.
[c]93.
[d]95.

Additional studies, using many more isolates from many geographical regions, are needed to reconcile these conflicting claims. It is important to determine whether the plasmids encode proteins that enable the parasite to thrive in the hostile environment of the phagolysosome, antibiotic resistance, or other virulence factors, such as adhesins. In addition to the genetic and biochemical differences described, the isolates also demonstrate different biological properties. For example, both *C. burnetii* Nine Mile (NM) isolate and *C. burnetii* Priscilla (PRS) isolate infected guinea pigs at does of two to four organisms per animal as determined by seroconversion. However, guinea pigs given 10 *C. burnetii* NM phase I organisms developed fever, while 10^6 or more *C. burnetii* PRS organisms was required to cause fever.[78] *C. burnetii* NM phase I kills most embryonated eggs by day 8 postinfection whereas equivalent numbers of *C. burnetii* PRS rarely kill the embryo. This altered virulence for embryonated eggs exhibited by *C. burnetii* PRS is seen in all isolates in Genomic Groups IV and V. All *C. burnetii* readily establish persistent infection in continuous cell lines where the organisms proliferate within phagolysosomal vacuoles.[1,58,95] Although persistently infected cells exhibit no differences in doubling time or cell cycle patterns, differences are seen in the growth rates of Genomic Groups I and IV in both L929 and J774 macrophage cells.[1,14,58,96] These data suggest that isolates differ in a number of ways which may be related to their ability to initiate chronic disease.

Whether or not there is a correlation between plasmid type and specific disease state does not exclude the distinct possibility that their plasmids encode virulence factors common to all *C. burnetii* variants essential for survival within the phagolysosome. It may be that *C. burnetii* plasmids carry genes critical for both acute and chronic disease, but that some of the genes are silent.

9. PROSPECTS

Much effort has been expended trying to identify *C. burnetii* virulence factors, including those that facilitate the survival of the parasite within the harsh environment of the phagolysosome. These studies, however, are hampered by the obligate intracellular nature of the organism. As shown in this review, molecular biological technology coupled with the development and use of host-cell model systems is facilitating the study and elucidation of the genetic makeup of the organism and the identification and characterization of its biological and biochemical properties. This technology has also revealed the presence in *C. burnetii* of a heat-shock operon that encodes a major antigen that is similar to a protein found in mycobacteria and *E. coli*[97] and a gene product that activates capsule synthesis in *E. coli*.[98] The latter observation is certainly of interest, although, to date, no capsule has been reported surrounding *C. burnetii*. The continued molecular analysis of *Coxiella* will help to resolve the roles in pathogenesis of factors that have already been identified, and will most certainly reveal other virulence determinants. Successful transformation of *C. burnetii* should eventually be attained with the aid of techniques such as electroporation, and this will be critical to future investigations.

That Q fever is a public health concern is underscored by recent retrospective surveys[99] that revealed, for the first time, the widespread occurrence of Q fever in Japan. The investigator's data indicate that prevalence of the agent is on the upswing and that chronic infection is a more common occurrence than had previously been recognized. Because of the poor prognosis associated with chronic disease—even with treatment—it becomes even more important to understand the agent's virulence determinants.

ACKNOWLEDGMENT. This work was partially supported by grants from the National Institutes of Health (NIAID) to O.G.B. (AI32492) and L.P.M. (AI20190).

REFERENCES

1. Baca, O. G., and Paretsky, D., 1983, Q fever and *Coxiella burnetii:* A model for host–parasite interactions, *Microbiol. Rev.* **47:**127–149.
2. Derrick, E. H., 1973, The course of infection with *Coxiella burnetii, Med. J. Aust.* **1:**1051–1057.
3. Marmion, B. P., 1962, Subacute rickettsial endocarditis: An unusual complication of Q fever, *J. Hyg. Epidemiol. Microbiol. Immunol.* **6:**79–84.
4. Baca, O. G., Klassen, D. A., and Aragon, A. S., 1993, Entry of *Coxiella burnetii* into host cells, *Acta Virol.* **37:**143–155.
5. Akporiaye, E. T., Rowatt, J. D., Aragon, A. S., and Baca, O. G., 1983, Lysosomal response of a murine macrophage-like cell line persistently infected with *Coxiella burnetii, Infect. Immun.* **40:**1155–1162.
6. Burton, P. R., Kordova, N., and Paretsky, D., 1971, Electron microscopic studies of the rickettsia *Coxiella burnetii:* Entry, lysosomal response, and fate of rickettsial DNA in L-cells, *Can. J. Microbiol.* **17:**143–150.
7. Heinzen, R. A., Scidmore, M. A., Rockey, D. D., and Hackstadt, T., 1996, Differential interaction with endocytic and exocytic pathways distinguish parasitophorous vacuoles of *Coxiella burnetii* and *Chlamydia trachomatis, Infect. Immun.* **64:**796–809.
8. Roman, M. P., Coriz, P., and Baca, O., 1986, A proposed model to explain persistent infection of host cells with *Coxiella burnetii, J. Gen. Microbiol.* **132:**1415–1422.
9. Hechemy, K. E., McKee, M., Marko, M., Samsonoff, W., Roman, M., and Baca, O., 1993, Three dimensional reconstruction of *Coxiella burnetii*-infected cells by high voltage electron microscopy, *Infect. Immun.* **61:**4485–4488.
10. Hackstadt, T., and Williams, J. C., 1981, Biochemical stratagem for obligate parasitism of eukaryotic cells by *Coxiella burnetii, Proc. Natl. Acad. Sci. USA* **78:**3240–3244.
11. Baca, O. G., Akporiaye, E. T., and Rowatt, J. D., 1984, Possible biochemical adaptations of *Coxiella burnetii* for survival within phagocytes: Effect of antibody, in: *Microbiology—1984* (L. Leive and D. Schlessinger, eds.), American Society for Microbiology, Washington, DC, pp. 269–272.
12. Aragon, A. S., Pereira, H. A., and Baca, O., 1995, A cationic antimicrobial peptide enhances the infectivity of *Coxiella burnetii, Acta Virol.* **39:**223–226.
13. Hendrix, L. R., Samuel, J. E., and Mallavia, L. P., 1991, Differentiation of *Coxiella burnetii* isolates by restriction endonuclease-digested DNA separated on SDS-PAGE, *J. Gen. Microbiol.* **137:**269–276.
14. Mallavia, L. P., 1991, Genetics of rickettsiae, *Eur. J. Epidemiol.* **7:**213–221.
15. Samuel, J. E., Frazier, M. E., and Mallavia, L. P., 1985, Correlation of plasmid type and disease caused by *Coxiella burnetii, Infect. Immun.* **49:**775–779.
16. Beaman, L., and Beaman, B. L., 1984, The role of oxygen and its derivatives in microbial pathogenesis, *Annu. Rev. Microbiol.* **38:**27–48.
17. Akporiaye, E. T., and Baca, O. G., 1983, Superoxide anion production and superoxide dismutase and catalase activities in *Coxiella burnetii, J. Bacteriol.* **154:**520–523.
18. Heinzen, R. A., Frazier, M. E., and Mallavia, L. P., 1992, *Coxiella burnetii* superoxide dismutase gene: Cloning, sequencing, and expression in *Escherichia coli, Infect. Immun.* **60:**3814–3823.
19. Hassan, H. M., and Fridovich, I., 1979, Paraquat and *Escherichia coli, J. Biol. Chem.* **254:**10846–10852.
20. Chary, P., Dillon, D., Schroeder, A. L., and Natvig, D. O., 1994, Superoxide dismutase (sod-1) null mutants of *Neurospora crassa:* Oxidative stress sensitivity, spontaneous mutation rate and response to mutagens, *Genetics* **137:**1–8.
21. Gralla, E. B., and Valentine, J. S., 1991, Null mutants of *Saccharomyces cerevisiae* Cu, Zn superoxide dismutase: Characterization and spontaneous mutation rates, *J. Bacteriol.* **173:**5918–5920.

22. Halliwell, B., and Gutteridge, J. M., 1990, Role of free radicals and catalytic metal ions in human disease: An overview, *Methods Enzymol.* **186:**1–85.
23. Akporiaye, E. T., Stefanovich, D., Tsosie, V., and Baca, O., 1990, *Coxiella burnetii* fails to stimulate human neutrophil superoxide anion production, *Acta Virol.* **34:**64–70.
24. Ferencik, M., Schramek, Kazar, J., and Stefanovic, J., 1984, Effect of *Coxiella burnetii* on the stimulation of hexose monophosphate shunt and on superoxide anion production in human polymorphonuclear leukocytes, *Acta Viol.* **28:**246–250.
25. Das, S., Saha, A. K., Remaley, A. T., Glew, R. H., Dowling, J. N., Kajiyoshi, M., and Gottlieb, M., 1986, Hydrolysis of phosphoproteins and inositol phosphates by cell surface phosphatase of *Leishmania donovani*, *Mol. Biochem. Parasitol.* **20:**143–153.
26. Saha, A. K., Dowling, J. N., Pasculle, A. W., and Glew, R. H., 1988, *Legionella micdadei* phosphatase catalyzes the hydrolysis of phosphatidylinositol 4,5-bisphosphate in human neutrophils, *Arch. Biochem. Biophys.* **265:**94–104.
27. Baca, O. G., Roman, M. J., Glew, R. H., Christner, R. F., Buhler, J. E., and Aragon, A. S., 1993, Acid phosphatase activity in *Coxiella burnetii*: A possible virulence factor, *Infect. Immun.* **61:**4232–4239.
28. Bliska, J. B., and Black, D. S., 1995, Inhibition of the Fc receptor-mediated oxidative burst in macrophages by the *Yersinia pseudotuberculosis* tyrosine phosphatase, *Infect. Immun.* **63:**681–685.
29. Groisman, E. A., Chiao, E., Lipps, C. J., and Heffron, F., 1989, *Salmonella typhimurium phoP* virulence gene is a transcriptional regulator, *Proc. Natl. Acad. Sci. USA* **86:**7077–7081.
30. Miller, S. I., Kukral, A. M., and Mekalanos, J. J., 1989, A two-component regulatory system (*phoP phoQ*) controls *Salmonella typhimurium* virulence, *Proc. Natl. Acad. Sci. USA* **86:**5054–5058.
31. Stock, J. B., Ninfa, A. J., and Stock, A. M., 1989, Protein phosphorylation and regulation of adaptive responses in bacteria, *Microbiol. Rev.* **53:**450–490.
32. Fields, P. I., Groisman, E. A., and Heffron, F., 1989, A *Salmonella* locus that controls resistance to microbiocidal proteins from phagocytic cells, *Science* **243:**1059–1062.
33. Mo, Y.-Y., and Mallavia, L. P., 1994, A *Coxiella burnetii* gene encodes a sensor-like protein, *Gene* **151:**185–190.
34. Cianciotto, N. P., Eisenstein, B. I., Mody, C. H., Toews, G. B., and Engleberg, N. C., 1989, A *Legionella pneumophila* gene encoding a species-specific surface protein potentiates initiation of intracellular infection, *Proc. Natl. Acad. Sci. USA* **89:**5188–5191.
35. Bangsborg, J. M., Cianciotto, N. P., and Hindersson, P., 1991, Nucleotide sequence analysis of the *Legionella micdadei mip* gene, encoding a 30-kilodalton analog of the *Legionella pneumophila* Mip protein, *Infect. Immun.* **59:**3836–3840.
36. Cianciotto, N. P., Bangsborg, J. M., Eisenstein, B. I., and Engleberg, N. C., 1990, Identification of *mip*-like genes in the genus *Legionella*, *Infect. Immun.* **58:**2912–2918.
37. Fischer, G., Bang, H., Ludwig, B., Mann, K., and Hacker, J., 1992, Mip protein of *Legionella pneumophila* exhibits peptidyl-prolyl-*cis*/*trans* isomerase (PPIase) activity, *Mol. Microbiol.* **6:**1375–1383.
38. Lundemose, A., G., Rouch, D. A., Birkelund, S., Christiansen, G., and Pearce, J. H., 1992, *Chlamydia trachomatis* Mip-like protein, *Mol. Microbiol.* **6:**2539–2548.
39. Cianciotto, N. P., and Fields, B. S., 1992, *Legionella pneumophila mip* gene potentiates intracellular infection of protozoa and human macrophages, *Proc. Natl. Acad. Sci. USA* **89:**5188–5191.
40. Cianciotto, N. P., Kim-Stamos, J. K., and Kamp, D. W., 1995, Infectivity of a *Legionella pneumophila mip* mutant for alveolar epithelial cells, *Curr. Microbiol.* **30:**247–250.
41. Cianciotto, N. P., O'Connell, W., Dasch, G. A., and Mallavia, L. P., 1995, Detection of *mip*-like sequences and Mip-related proteins within the family *Rickettsiaceae*, *Curr. Microbiol.* **30:**149–153.
42. Mo, Y.-Y., Cianciotto, N. P., and Mallavia, L. P., 1995, A *Coxiella burnetii* Mip analog, *Microbiology* **141:**2861–2871.
43. Howe, D., Seshu, J., and Mallavia, L. P., 1995, Transmission electron microscopy localizes

Coxiella burnetii macrophage infectivity potentiator antigen to host cell cytoplasm, *Abstr. Annu. ASM Meet.* **D128:**271.

44. Hurley, M. C., Balaazovich, K., Albano, M., Engleberg, N. C., and Eisenstein, B. E., 1993, *Legionella pneumophila* Mip inhibits protein kinase C, in: *Legionella. Current Status and Emerging Perspectives* (J. M. Barbaree, R. F., Breiman, and A. P. Dufour, eds.), American Society for Microbiology, Washington, DC, pp. 69–70.

45. Lundemose, A. G., Kay, J. E., and Pearce, J. H., 1993, *Chlamydia trachomatis* Mip-like protein has peptidyl-prolyl *cis/trans* isomerase activity that is inhibited by FK506 and rapamycin and is implicated in initiation of chlamydial infection, *Mol. Microbiol.* **7:**777–783.

46. Bierer, B. E., Mattila, P. S., Standaert, R. F., Herzenberg, L. A., Burakoff, S. J., Crabtree, G., and Schreiber, S. L., 1990, Two distinct signal transmission pathways in T lymphocytes are inhibited by complexes formed between an immunophilin and either FK506 or rapamycin, *Proc. Natl. Acad. Sci. USA* **87:**9231–9235.

47. Dumont, F. J., Melino, M. R., Staruch, M. J., Koprak, S. L., Fischer, P. A., and Sigal, N. H., 1990, The immunosuppressive macrolides FK-506 and rapamycin act as reciprocal antagonists in murine T cells, *J. Immunol.* **144:**1418–1424.

48. Hacker, J., and Fischer, G., 1993, Immunophilins: Structure–function relationship and possible role in microbial pathogenicity, *Mol. Microbiol.* **10:**445–456.

49. Clipstone, N. A., and Crabtree, G. R., 1992, Identification of calcineurin as a key signalling enzyme in T-lymphocyte activation, *Nature* **357:**695–697.

50. Dumont, F. J., Staruch, M. J., Koprak, S. L., Melino, M. R., and Sigal, N. H., 1990, Distinct mechanisms of suppression of murine T cell activation by the related macrolides FK-506 and rapamycin, *J. Immunol.* **144:**251–258.

51. Hendrix, L., and Samuel, J. E., 1995, The outer membrane-associated protein disulfide oxido-reductase, Com-1, of *Coxiella burnetii*, *Abstr. Annu. ASM Meet.* **D131:**272.

52. Hendrix, L. R., Mallavia, L. P., and Samuel, J. E., 1993, Cloning and sequencing of *Coxiella burnetii* outer membrane protein gene *com1*, *Infect. Immun.* **61:**470–477.

53. von Heijne, G., 1985, Signal sequences, the limits of variation, *J. Mol. Biol.* **184:**99–105.

54. Gilbert, H. F., 1990, Molecular and cellular aspects of thiol-disulfide exchange, *Adv. Enzymol.* **63:**69–172.

55. Hawkins, H. C., Blackburn, E. C., and Freedman, R. B., 1991, Comparison of the activities of protein disulphide-isomerase and thioredoxin in catalysing disulphide isomerization in a protein substrate, *Biochem. J.* **275:**349–353.

56. Snyder, C. E., Jr., and Williams, J. C., 1986, Purification and chemical characterization of a major membrane protein from *Coxiella burnetii*, *Abstr. Annu. ASM Meet.* **K-1:**193.

57. Williams, J. C., Hoover, T. A., Waag, D. M., Banerjee-Bhatnagar, N., Bolt, C. R., and Scott, G. H., 1990, Antigenic structure of *Coxiella burnetii*: A comparison of lipopolysaccharide and protein antigens as vaccines against Q fever, *Ann. N.Y. Acad. Sci.* **590:**370–380.

58. Roman, M. J., Samuel, J. E., Mallavia, L. P., and Baca, O. G., 1987, Comparison of the morphologies and growth characteristics of cultured host cells exposed to the Nine Mile and Priscilla isolates of *Coxiella burnetii*, *Abstr. Annu. ASM Meet.* **B263:**68.

59. Schmeer, N., 1988, Early recognition of a 27-kDa membrane protein (MP27) in *Coxiella burnetii* infected and vaccinated guinea pigs, *J. Vet. Med. Ser. B* **35:**338–345.

60. Schmeer, N., Muller, H.-P., Baumgartner, W., Wieda, J., and Krauss, H., 1988, Enzyme-linked immunosorbent fluorescence assay and high-pressure liquid chromatography for analysis of humoral immune responses to *Coxiella burnetii* proteins, *J. Gen. Microbiol.* **26:**2520–2525.

61. Amano, K. I., and Williams, J. C., 1984, Chemical and immunological characterization of lipopolysaccharides from phase I and phase II *Coxiella burnetii*, *J. Bacteriol.* **160:**994–1002.

62. Baca, O. G., and Paretsky, D., 1974, Partial chemical characterization of a toxic lipopolysaccharide from *Coxiella burnetii*, *Infect. Immun.* **9:**959–961.

63. Schramek, S., and Brezina, R., 1976, Characterization of an endotoxic lipopolysaccharide from *Coxiella burnetii*, *Acta Virol.* **20:**152–158.

64. Hackstadt, T., 1986, Antigenic variation in the phase I lipopolysaccharide of *Coxiella burnetii* isolates, *Infect. Immun.* **52:**337–340.

65. Hackstadt, T., Peacock, M. G., Hitchcock, P. J., and Cole, R. L., 1985, Lipopolysaccharide variation in *Coxiella burnetii:* Intrastrain heterogeneity in structure and antigenicity, *Infect. Immun.* **48:**359–365.

66. Schramek, S., and Mayer, H., 1982, Different sugar compositions of lipopolysaccharides isolated from phase I and pure phase II cells of *Coxiella burnetii*, *Infect. Immun.* **38:**53–57.

67. Baca, O. G., Martinez, I. L., Aragon, A. S., and Klassen, D., 1980, Isolation and partial characterization of a lipopolysaccharide from phase II *Coxiella burnetii*, *Can. J. Microbiol.* **26:**819–826.

68. Stendahl, O., and Edebo, L., 1972, Phagocytosis of mutants of *Salmonella typhimurium* by rabbit polymorphonuclear cells, *Acta Pathol. Microbiol. Scand.* **B80:**481–488.

69. Baca, O. G., Akporiaye, E. T., Aragon, A. S., Martinez, I. L., Robles, M. V., and Warner, N. L., 1981, Fate of phase I and phase II *Coxiella burnetii* in several macrophage-like tumor cell lines, *Infect. Immun.* **33:**258–266.

70. Vishwanath, S., and Hackstadt, T., 1988, Lipopolysaccharide phase variation determines the complement-mediated serum susceptibility of *Coxiella burnetii*, *Infect. Immun.* **56:**40–44.

71. Roantree, R. J., and Rantz, L. A., 1960, A study of the relationship of the normal bactericidal activity of human serum to bacterial infection, *J. Clin. Invest.* **39:**72–81.

72. Hackstadt, T., 1988, Steric hindrance of antibody binding to surface proteins of *Coxiella burnetii* by phase I lipopolysaccharide, *Infect. Immun.* **56:**802–807.

73. Baca, O. G., and Paretsky, D., 1974, Some physiological and biochemical effects of a *Coxiella burnetii* lipopolysaccharide preparation on guinea pigs, *Infect. Immun.* **9:**939–945.

74. Schramek, S., Kazar, J., Sekeyova, Z., Freudenberg, M. A., and Galanos, C., 1984, Induction of hyperreactivity to endotoxin in mice by *Coxiella burnetii*, *Infect. Immun.* **45:**713–717.

75. Kimbrough, R. C., III, Ormsbee, R. A., and Peacock, M. G., 1981, Q fever endocarditis: A three and one-half year follow-up, in: *Rickettsia and Rickettsial Diseases* (W. Burgdorfer and R. L. Anacker, eds.), Academic Press, New York, pp. 125–132.

76. Peacock, M. G., Philip, R. N., Williams, J. C., and Faulkner, R. S., 1983, Serological evaluation of Q fever in humans: Enhanced phase I titers of immunoglobulins G and A are diagnostic for Q fever endocarditis, *Infect. Immun.* **41:**1089–1098.

77. Tobin, M. J., Cahill, N., Gearty, G., Maurer, B., Blake, S., Daly, K., and Hone, R., 1982. Q fever endocarditis, *Am. J. Med.* **72:**396–400.

78. Moos, A., and Hackstadt, T., 1987, Comparative virulence of intra- and interstrain lipopolysaccharide variants of *Coxiella burnetii* in the guinea pig model, *Infect. Immun.* **55:**1144–1150.

79. Samuel, J. E., Frazier, M. E., Kahn, M., Thomashow, L. L. S., and Mallavia, L. P., 1983, Isolation and characterization of a plasmid from phase I *Coxiella burnetii*, *Infect. Immun.* **41:**488–493.

80. Savinella, E. A., and Mallavia, L. P., 1990, Comparison of the plasmids QpRS and QpH1 to homologous sequences present in a plasmidless endocarditis-causing isolate of *Coxiella burnetii*, *Ann. N.Y. Acad. Sci.* **590:**523–533.

81. Frazier, M. E., Heinzen, R. A., Stiegler, G., and Mallavia, L. P., 1991, Physical mapping of the *Coxiella burnetii* genome, *Acta Virol.* **35:**511–518.

82. Heinzen, R. A., Stiegler, G. L., Whiting, L. L., Schmitt, S. A., Mallavia, L. P., and Frazier, M. E., 1990, Use of pulsed field electrophoresis to differentiate *Coxiella burnetii* strains, *Ann. N.Y. Acad. Sci.* **590:**504–513.

83. Mallavia, L. P., Samuel, J. E., and Frazier, M. E., 1991, The genetics of *Coxiella burnetii*, etiologic agent of Q fever and chronic endocarditis, in: *Q Fever: The Biology of Coxiella burnetii* (J. C. Williams and H. Thompson, eds.), CRC Press, Boca Raton, pp. 259–284.

84. Yeaman, M. R., and Baca, O. G., 1990, Antibiotic susceptibility of *Coxiella burnetii*, in: *Q Fever, Volume 2* (T. Marie, ed.), CRC Press, Boca Raton, pp. 213–223.

85. Lin, Z., and Mallavia, L. P., 1994, Identification of a partition region carried by the plasmid QpH1 of *Coxiella burnetii*, *Mol. Microbiol.* **13:**513–523.

86. Lin, Z., and Mallavia, L. P., 1995, The partition region of plasmid QpH1 is a member of a family of two *trans*-acting factors as implied by sequence analysis, *Gene* **161:**69–74.

87. Lin, Z., Howe, D., and Mallavia, L. P., 1995, Roa307, a protein encoded on *Coxiella burnetii* plasmid QpH1, shows homology to proteins encoded in the replication origin region of bacterial chromosomes, *Mol. Gen. Genet.* **243:**487–490.

88. Minnick, M. F., Small, C. L., Frazier, M. E., and Mallavia, L. P., 1991, Analysis of the cbhE' plasmid gene from acute disease-causing isolates of *Coxiella burnetii*, *Gene* **103:**113–118.

89. Minnick, M. F., Heinzen, R. A., Reschke, D. K., Frazier, M. E., and Mallavia, L. P., 1991, A plasmid-encoded surface protein found in chronic-disease isolates of *Coxiella burnetii*, *Infect. Immun.* **59:**4735–4739.

90. Stein, A., and Raoult, D., 1993, Lack of pathotype specific gene in human *Coxiella burnetii* isolates, *Microb. Pathogen.* **15:**177–185.

91. Atzpodien, E., Baumgartner, W., Artelt, A., and Thiele, D., 1994, Valvular endocarditis occurs as a part of a disseminated *Coxiella burnetii* infection in immunocompromised BALB/cJ (H-2^d) mice infected with the Nine Mile isolate of *C. burnetii*, *J. Infect. Dis.* **170:**223–226.

92. Thiele, D., and Willems, H., 1994, Is plasmid based differentiation of *Coxiella burnetii* in 'acute' and 'chronic' isolates still valid? *Eur. J. Epidemiol.* **10:**427–434.

93. Valkova, D., and Kazar, J., 1995, A new plasmid (QpDV) common to *Coxiella burnetii* isolates associated with acute and chronic Q fever, *FEMS Microbiol. Lett.* **125:**275–280.

94. Thiele, D., Willems, H., Haas, M., and Krauss, H., 1994, Analysis of the entire nucleotide sequence of the cryptic plasmid QpH1 from *Coxiella burnetii*, *Eur. J. Epidemiol.* **10:**413–420.

95. Ning, Z., Shu-Rong, Y., Quan, Y. G., and Xue, Z., 1991, Molecular characterization of cloned variants of *Coxiella burnetii* isolated in China, *Acta Virol.* **35:**173–183.

96. Roman, M., Crissman, H., Samsonoff, W., Hechemy, K., and Baca, O., 1991, Analysis of *Coxiella burnetii* isolates in cell culture and the expression of parasite-specific antigens on the host membrane surface, *Acta Virol.* **35:**503–510.

97. Vodkin, M. H., and Williams, J. C., 1988, A heat shock operon in *Coxiella burnetii* produces a major antigen homologous to a protein in both mycobacteria and *Escherichia coli*, *J. Bacteriol.* **170:**1227–1234.

98. Zuber, M., Hoover, T. A., and Court, D. L., 1995, Analysis of *Coxiella burnetii* gene product that activates capsule synthesis in *Escherichia coli:* Requirement for the heat shock chaperone DnaK and the two-component regulator RcsC, *J. Bacteriol.* **177:**4238–4244.

99. Yuasa, Y., Yoshe, K., Takasaki, T., Yoshida, H., and Oda, H., 1996, Retrospective survey of chronic Q fever in Japan by using PCR to detect *Coxiella burnetii* DNA in paraffin-embedded clinical samples, *J. Clin. Microbiol.* **34:**824–827.

100. Li, Y. P., Curley, G., Lopez, M., Glew, R., Aragon, A., Kumar, H., and Baca, O. G., 1996, Protein-tyrosine phosphatase activity of *Coxiella burnetii* that inhibits human neutrophils, *Acta. Virol.* **40:**263–272.

9

Human Granulocytic Ehrlichiosis

J. STEPHEN DUMLER and PHILIPPE BROUQUI

1. INTRODUCTION

Ticks are largely feared for their ability to transmit agents of severe and fatal human disease. In recent years, a heightened fear of ticks has occurred because of the discovery of previously undescribed human illness caused by spirochetes, rickettsiae, and viruses that were unknown or suspected to cause disease in animals only. The recognition of the increased spectrum of potential pathogenic agents transmissible by tick vectors has also awakened scientists to the ecologic factors that lead to the emergence of these infections, the maintenance of the agents in nature, and the mechanisms by which these agents use tick and mammalian reservoirs for an evolutionary advantage and species maintenance. This chapter will describe the novel tick-borne human disease currently called human granulocytic ehrlichiosis (HGE) and will present pertinent data concerning the clinical disease and immunologic aspects relevant to the disease in humans and animals. The adaptations that allow for efficient transmission and persistance of the agent in nature provide an evolutionary survival scheme, and are the same as those that contribute to significant human and veterinary disease.

J. STEPHEN DUMLER • Division of Medical Microbiology, Department of Pathology, The Johns Hopkins Medical Institutions, Baltimore, Maryland 21287-7093.   PHILIPPE BROUQUI • Unitée des Rickettsies, CNRS-UPRES-A, Faculté de Médecine, 13385 Marseille cedex 5, France.

Rickettsial Infection and Immunity, edited by Anderson *et al.* Plenum Press, New York, 1997.

2. GRANULOCYTOTROPIC *EHRLICHIA* SPECIES: CLASSIFICATION AND TAXONOMIC CONSIDERATIONS

For many years, members of the genus *Ehrlichia* were categorized based solely on four features: infected host cell, infected mammalian species, geographic location, and antigenic cross-reactivity.[1] This situation existed because *in vitro* cultivation was possible for a few species only. Recent times have seen the development of molecular tools for phylogenetic studies and improved methods for cultivation that have helped to clarify the exact phylogenetic positions of these obligate intracellular bacteria.[2]

Ehrlichiae infect predominantly cells of bone marrow origin *in vivo*.[13] These bacteria have a typical gram-negative-type cell wall, and contain ribosomes, RNA, and DNA. Lipopolysaccharide and lipo-oligosaccharides are in minute concentrations, if present at all. Morphologically, the ehrlichiae live within a membrane-bound vacuole, shown to be a phagosome for two species currently classified as *Ehrlichia*.[1] Extrapolation from the model of *E. risticii* would indicate that these ehrlichiae attach to a cell surface protein via a bacterial protease-sensitive adhesin.[4] The attached ehrlichiae inhibit phagosome–lysosome fusion by active protein synthesis.[5] The phagosome–lysosome fusion inhibition can be abrogated by gamma-interferon or an increased intracellular Ca^{2+} concentration such as is seen with calcium ionophores.[6] The intraphagosomal ehrlichiae divide by binary fission to produce a cluster called a morula. Multiple morulae may be seen in one cell; thus, phagocytic vacuoles apparently do not fuse.

Currently, *Ehrlichia* species are divided into genogroups based on the nucleic acid sequence of the 16 S ribosomal RNA gene of each.[3,7] The three genogroups are the *Ehrlichia canis* group, the *E. sennetsu* group, and the *E. phagocytophila* group. The latter group includes two named species and one unnamed species that are granulocytotropic *in vivo* and bear remarkable genetic, antigenic, and biological similarities.[2,8–11] *E. phagocytophila* has long been known as the agent of tick-borne fever, a febrile disease of ruminants mostly in Europe.[1] *E. equi* is the agent of equine granulocytic ehrlichiosis, recognized first in California, but now increasingly seen worldwide. The unnamed agent, the human granulocytic ehrlichia, shares 99.9 and 99.8% sequence similarity with *E. phagocytophila* and *E. equi*, respectively, and antigenic analyses indicate the presence of strong serologic cross-reactivity by indirect fluorescent antibody tests and immunoblot analysis using *E. equi* as antigen.[8,9]

Both *E. equi* and the human granulocytic ehrlichia have been cultivated *in vitro* using human promyelocyte cell lines and tick cell lines, where each retains antigenic reactivity with polyclonal antisera from experimental and natural infections.[12,13] Unlike *E. chaffeensis* and *E. canis*, rapid isolation of these granulocytotrophic ehrlichiae is possible especially if large numbers of organisms are present in the blood inoculum.[14]

These granulocytotropic ehrlichiae contain at least two highly conserved genes, a 16 S rRNA gene and a *gro*EL heat shock/chaperonin gene-equivalent.[2,8,15] Slightly more divergence in gene sequence of the *gro*EL operon is present between those agents identified in North America (*E. equi* and the HGE agent) and in Europe (*E. phagocytophila*).[15] Aside from these genes, no other genes with known function have been identified in this genogroup. Immunoblot analyses indicate the presence of several protein antigens that react predominantly or exclusively with antisera to *E. phagocytophila*-group agents and not other *Ehrlichia* or closely related bacterial species.[9] This includes a dominant 44-kDa antigen reaction, and minor reaction with 25-, 42-, and 100-kDa proteins. These ehrlichiae also contain a variety of protein antigens that range between 56 and 80 kDa and have broad serologic cross-reactivity with antisera to other *Ehrlichia* species, *Rickettsia* species, and other bacteria. These bands presumably represent heat-shock proteins that may include the products of the *gro*EL operon.

3. HUMAN GRANULOCYTIC EHRLICHIOSIS: CLINICAL CONSIDERATIONS

Ehrlichioses in humans were known to be caused by only two agents in 1990: *E. sennetsu*, which had been identified only in the Far East, and *E. chaffeensis*, a pathogen of primarily mononuclear phagocytes *in vivo*, which was recognized mostly in the south central and southeastern United States.[2,16] In the summer of 1990, a patient died after an apparent ehrlichia infection during which only neutrophils could be demonstrated to harbor the infectious agent in peripheral blood.[8,17] This distinctive finding, the lack of antibodies in convalescent serum from additional patients that reacted with *E. canis* or *E. chaffeensis*, and the lack of *E. chaffeensis* DNA in the blood of these patients suggested an alternative explanation: an infection with a previously unknown human ehrlichial pathogen. Subsequent molecular studies identified the new human granulocytic ehrlichia in most of the blood samples from these patients and showed serologic reactivity with both *E. equi* and *E. phagocytophila*. Simultaneously, similar infections were recognized in dogs and horses in overlapping and remote geographic areas, including the upper Midwest,[18] the Northeast,[19] and northern California[20] in the United States and Sweden in Europe.[21] Many human patients reported deer tick *Ixodes scapularis*) bites, and several patients also developed antibody responses to both *Borrelia burgdorferi* and *Babesia microti*, which are also transmitted by deer ticks.[22–24]

Recently, serologic investigations demonstrated that 3 to 17% of patients who were bitten by *Ixodes* ticks in Europe (Italy, Sweden, Switzerland)[25] had significant antibody titers to *E. equi*. Moreover, *E. phagocytophila* 16 S rRNA gene sequences have been found in 2% of engorged *Ixodes* ticks in France.

The clinical illness associated with infection by the HGE agent is very similar to that described for *E. chaffeensis*, and is mostly characterized by fever, headache, malaise, and myalgias.[26] Other findings occur in less than half of patients, but may be important clues to aid in a clinical or presumptive diagnosis including anorexia, nausea, vomiting, diarrhea, cough, and, occasionally, confusion or other evidence of potential CNS involvement. Rash is a very infrequent finding in HGE, seen in less than 2% of patients. Although clinical laboratory findings are also nonspecific, the clinical findings coupled with the presence of leukopenia, thrombocytopenia, anemia, and elevations in hepatic transaminase activities are very suggestive of the ehrlichioses.[3,7] Diagnosis is usually made retrospectively by serologic means; however, a specific diagnosis may be achieved at the time of illness by PCR amplification of ehrlichial DNA from acute-phase blood samples, preferably obtained prior to doxycycline therapy.[27] Occasionally, review of the Wright-stained peripheral blood smear will reveal the presence of morulae in leukocytes, usually neutrophils, for HGE. This finding is rare in monocytic ehrlichiosis caused by *E. chaffeensis*, but may be seen as often as in one in five patients with HGE.[3,26]

HGE is usually an uncomplicated febrile illness that is easily and rapidly treated with a tetracycline antibiotic such as doxycycline. Most patients respond with a rapid defervescence and resolution of clinical manifestations within 48 to 72 hr. However, serious and even fatal infections have been documented to occur, including respiratory distress, hemorrhage, shock, and renal failure.[26] The few fatalities that have occurred in association with HGE apparently resulted from opportunistic and nosocomial infections, and strongly suggest that primary HGE causes some abnormality in host defense or immune function.

4. IMMUNE REACTIONS IN RECOVERED HOSTS

Humans who acquire HGE develop serum IgG and IgM antibodies that react with the HGE agent in horse neutrophils[10] and in human promyelocyte HL60 cells,[12] with *E. equi* in equine neutrophils, and with *E. phagocytophila* in bovine neutrophils.[17] The development of these antibodies further supports the close phylogenetic positions of these granulocytic ehrlichiae. The occurrence of serologic cross-reactions in humans was further investigated by means of immunoblots using *E. equi* purified by density gradient centrifugation from the blood of infected horses[9] (Fig. 1). By this method, it was shown that humans convalescent from HGE, dogs after canine *"E. equi"* infection, horses after *E. equi* infection, and cattle after tick-borne fever (*E. phagocytophila* infection) all develop antibodies that react with a protein antigen of approximately 44 kDa. Antisera from humans and animal species convalescent from other natural ehrlichial, rickettsial, and bacterial infections, or immunized with other *Ehrlichia* species, rickettsiae, or bacteria

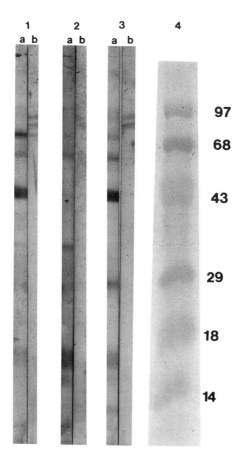

FIGURE 1. *Ehrlichia equi* immunoblot for demonstration of antibodies in human granulocytic ehrlichiosis. The *E. equi* were purified from infected horse whole blood leukocytes by lysis and renografin density gradient centrifugation before SDS-PAGE and immunoblot electrotransfer. Lanes 1 and 3 were reacted with serum from two patients convalescent from human granulocytic ehrlichiosis; lane 2 was reacted with serum from a normal, healthy volunteer with no history of HGE. Lanes labeled "a" contain *E. equi* antigen; lanes labeled "b" contain lysed normal horse leukocyte antigen as control. Note the dense band at approximately 44 kDa in lanes 1a and 3a that is not present in other lanes. Lane 4 contains the Coomassie blue-stained molecular size markers that are shown in kDa on the right. Bound human antibodies were detected with a biotinylated goat poly-valent (IgG, IgA, and IgM) anti-human immunoglobulin, followed by reaction with alkaline phospha-tase-labeled streptavidin, and BCIP/NBT as chromogen.

did not react with this antigen. Some granulocytic ehrlichia antisera reacted with other antigens of *E. equi* not detected by antisera to other bacteria, including 25-, 42-, and 100-kDa antigens. Importantly, frequent serologic reactions were observed with *E. equi* antigens broadly ranging between 56 and 75 kDa, probably representing *gro*EL and other heat-shock proteins.

At the time of clinical presentation, less than 25% of patients with HGE will have antibodies detected by IFA serology using *E. equi*-infected leukocytes as antigen. However, nearly all patients will develop titers in excess of 80 within the first month, and many will still have antibodies detectable for at least 3 years after primary infection.[26] Not all infections with the HGE agent result in the production of antibodies detectable by the routine IFA method using *E. equi* as antigen, and it is suspected that early therapy may abrogate antigenic mass sufficiently to preclude a detectable antibody response. Serologic reactions with *E. chaffeensis* antigens have been demonstrated after HGE. These reactions are present in only a small proportion of all HGE patients tested by current methods and suggest that the causative ehrlichiae share some antigenic determinants.[26,28]

No studies of cell-mediated immunity to the HGE agent have been performed. However, when blood from a human patient with HGE was transfused into a horse, signs and laboratory findings typical for equine granulocytic ehrlichiosis developed; after convalescence, the horse was immune to challenge with viable *E. equi*.[11,19] This finding is a strong indication that cross-immunity between the HGE agent and *E. equi* can be established, or that these two agents are biological equivalents.

5. IMMUNE FUNCTION IN HUMAN PATIENTS WITH ACUTE HGE

The majority of patients with HGE develop normal humoral immunity within 1 month of onset of illness, and the vast majority of identified patients recover with or without doxycycline therapy.[26] However, 7% of patients have severe enough illness to warrant admission to an intensive care unit, and the case fatality rate is approximately 5%. Postmortem examinations revealed that of four fatalities, three died of opportunistic infections, and the fourth died of a fatal cardiac arrhythmia after myocarditis.[26,28] Two of the fatalities exsanguinated from esophageal ulcers caused by a *Candida* species and herpesvirus infection, respectively, and a third died of severe pulmonary damage caused by invasive pulmonary aspergillosis.[29] In addition, the patient with herpesvirus esophagitis had cryptococcal pneumonia. Although two of these patients had potential reasons for immune compromise, each was considered in good health at the time of acute HGE, suggesting that the ehrlichia infection contributed to the defects in host defense and cell-mediated immune function that led to the opportunistic infections.

6. IMMUNITY IN ANIMAL MODELS OF GRANULOCYTOTROPHIC EHRLICHIOSES

The clinical, laboratory, and epidemiological similarities among tick-borne fever (*E. phagocytophila* infection), EGE (*E. equi* infection), and HGE, as well as the close genetic, antigenic, and biologic similarities in the causative agents indicate that data about veterinary infections may be relevant to human illness. The accrued data suggest that acute infections by *E. phagocytophila*-group *Ehrlichia* spp. cause immune suppression, dysfunctional host defenses, and persistent infections. In some cases, the consequences of simultaneous infection with one or more other pathogens lead to significant morbidity and mortality in the animals.

7. VETERINARY MODELS OF *EHRLICHIA PHAGOCYTOPHILA-*GROUP INFECTION

E. phagocytophila is a tick-borne, granulocytic ehrlichia known to infect goats, sheep, cattle, and deer in Europe.[1,30,31] The disease, called tick-borne fever (TBF), is characterized by morulae in circulating granulocytes[32] with an ultrastructural morphology identical with other species of *Ehrlichia*.[1,33] Infection is transmitted by nymphal or adult *I. ricinus* ticks, and transovarian transmission has not been demonstrated.[34] The infection in sheep occurs after a 3- to 13-day incubation period.[34,35] Fever ensues, the animals appear dull and listless, lose body weight, develop thrombocytopenia and leukopenia,[30,35-37] and have increasing numbers of infected neutrophils.[35] The illness resolves after 1 to 12 days, but relapse occasionally occurs. Splenectomy and other inflammatory disturbances have resulted in recrudescence up to 1 year after infection.[32] TBF predisposes frequently to severe concurrent bacterial, fungal, and viral infections and hemorrhages in sheep.[32,35,38-47] TBF is a major determinant in the development of fatal staphylococcal pyemia,[32,42,43] resulting in the death of 1–2% of the sheep population in Great Britain per year.[41] Although TBF is usually benign, the mortality rate has been as high as 24%.[41] TBF is easily treated with tetracyclines.[41]

Equine granulocytic ehrlichiosis (EGE), caused by *E. equi* and potentially the HGE agent, was first observed in California[48,49] and has now been diagnosed in Colorado, Illinois, Florida, Washington, Connecticut, and New Jersey in the United States,[50,51] and also in British Colombia in Canada,[52] Germany, Switzerland, Sweden, Norway,[53] Great Britain,[54] Wales,[55] and Venezuela.[56] The disease is characterized by fever, anorexia, depression, limb edema, jaundice, petechiae, and ataxia in infected horses.[20,48] Morulae with light microscopic and ultrastructural characteristics typical of the genus are observed in circulating neutrophils during the acute febrile phase.[10,48,57] Laboratory findings include leukopenia, thrombocytopenia, anemia, hyperbilirubinemia, and morulae in approximately 40% of cells.[20,48] Horses become afebrile within 24–48 hr after tetracycline therapy[20] and fatalities are unusual.[20,48] Severe secondary bacterial infections were seen in 4 of 31 experimentally infected horses.[48] *Ixodes pacificus* ticks are now known to be competent vectors for experimental transmission to horses.[58] EGE occurs during the fall, winter, and spring months in California,[20] concurrent with the activity of *I. pacificus* ticks in this geographic location. Seroepidemiologic surveys of horses in northern California have shown between 3.1 and 10.3% prevalence, and >50% seroprevalence is seen among horses on ranches enzootic for *E. equi*.[50] EGE caused by a granulocytic ehrlichia with a 16 S rRNA gene sequence identical to that described for the HGE agent has been identified in Connecticut[19] and Minnesota[18] in the United States, and in Sweden[21] as well.

Canine infections with *E. equi* or a related ehrlichia are also well documented. First isolated in horses inoculated with infected canine blood harboring granulocytic morulae,[59] the infection has been widely documented in domestic dog populations throughout the United States and probably worldwide, including 17 dogs from Minnesota and Wisconsin and 4 dogs from Sweden that were infected with an agent indistinguishable from the agent of HGE.[18,21,60,61]

The high degree of genetic and antigenic similarities among the members of the *E. phagocytophila* genetic group strongly suggest a very close relationship and potential biological identity. If this is so, a substantial human population base is at risk for infection, given the wide geographic distribution of these agents. A measure of the biological diversity of ehrlichiae in this genetic group is reflected in studies of infection by "isolates" of *E. phagoyctophila* from different geographic regions of Great Britain or in different mammalian host species. Results showed that predictable clinical illnesses resulted, but protection from reinfection with the homologous, but not heterologous, strains(s) could be demonstrated[32,62] indicating the presence of some biologically divergent strains.

Some host specificity or adaptation occurs, as experimental infections of other mammals with *E. phagocytophila*[32,44,63] or *E. equi*[49,64] had limited success. Foggie and Hood[63] laboriously (>100 passages) adapted *E. phagocytophila* in both splenectomized mice and guinea pigs; however, the clinical illness was always mild or inapparent. Similarly, inoculation of *E. equi*-infected horse blood into goats, lambs, cats, macaques, baboons, and dogs yielded mild or asymptomatic infections with low-level ehrlichemia. Infection of mice and rabbits was unsuccessful.[64] Recent studies have shown that inoculation of blood from patients with HGE into susceptible horses results in typical EGE, and that after convalescence, these HGE animals are protected against challenge by viable *E. equi*.[11,19]

One interpretation of this evidence suggests that the HGE agent, *E. equi*, and perhaps *E. phagocytophila* represent a single species with strains that have adapted to one or more mammalian hosts. In fact, as compared with *E. equi* and *E. phagocytophila*, the few unique nucleotide changes observed in the 16 S rRNA gene amplified from the first defined human case have been identified in at least eight additional human patients,[65] in three dogs from the upper Midwest,[18] in two horses from Connecticut[19] and a horse from California,[12] and in three horses and four dogs from Sweden.[21] Moreover, HGE agent-like DNA can be amplified from normal rodents from Minnesota and normal deer from Wisconsin and California, and sequence data show similar, but distinct nucleotide profiles suggestive of closely related or identical ehrlichiae.

8. IMMUNOBIOLOGY AND ANTIGENIC COMPARISONS OF HGE, *E. PHAGOCYTOPHILA*, AND *E. EQUI*

It is well recognized that, as in HGE,[17] *E. phagocytophila*[41] and *E. equi*[56] predispose to secondary infections. Moreover, underlying immune compromise may be a risk factor for development of more severe infection, as with *E. equi* in two dogs with underlying immune compromising illnesses[59] and in human fatalities.[17,26]

Investigations into the mechanisms that permit the secondary infectious complications after *E. phagocytophila* infection have revealed several potential explanations. Early in the infectious process, lymphopenia occurs in association with compromise of both B and T cell function.[46,66-68] Antibody production is diminished during *E. phagocytophila* infection, and although delayed-type hypersensitivity reactions remain intact,[69] impaired T-cell proliferative responses[67] and decreases in CD4+, CD8+, and CD5+ T-cell populations occur.[70] Serum obtained from experimentally infected sheep 3 weeks into convalescence contains a factor that inhibits lymphocyte proliferation in response to phytohemagglutinin, pokeweed mitogen, and concanavalin A.[68] In addition, infected neutrophils are defective at emigration and phagocytosis.[71,72] Abnormalities in these host defenses are permissive of opportunistic infections in man. If a similar situation occurs in humans with acute ehrlichiosis, the potential risk for similar dysfunction of host defenses and the development of opportunistic infections exists.

9. IMMUNITY AND PERSISTENT INFECTIONS BY *E. PHAGOCYTOPHILA*-GROUP EHRLICHIAE

Persistent infection of mammals by obligate intracellular bacteria is a well-recognized phenomenon. Recovery from infection is often associated with the development of immunity. However, human infection by the agent of HGE has resulted in prolonged illness with persistent infection, even when high titers of serum antibody are detected.[73] Regardless, most human patients recover without sequelae and develop high titers of antibody. Although reported in animals, recurrent or recrudescent infections have not yet been reported in humans.

Goats experimentally infected with *E. phagocytophila* also normally recover[32,34] and are immune.[74,75] This immunity is nonsterile, as animals remain infectious for up to 2 years.[32] In contrast, *E. equi* humoral and cellular immune responses are detectable for 300 and 200 days postinfection, respectively, but protective immunity may be incomplete.[76] Blood from tetracycline-treated horses recovered from ehrlichiosis is noninfectious.[20] Whether human infection results in immunity to reinfection is not certain; however, horses convalescent from HGE agent infection are protected from *E. equi* challenge.[11]

The epidemiologic and ecologic similarities among Lyme borreliosis and the zoonotic granulocytic ehrlichioses suggest that the *E. phagocytophila*-group ehrlichiae may establish subclinical persistent infections in the major natural reservoir hosts, small mammals. In fact, Tyzzer described natural infection of *Microtus pennsylvanicus* by an agent morphologically similar to the HGE agent in Nantucket, MA, in the 1930s.[77] Recent experiments have shown that 50% of the laboratory mice that were experimentally infected with *E. equi* maintained a subclinical infection for more than 50 days (unpublished data). These findings confirm the occasional finding of long-term infections or infectivity in animals that are highly susceptible to symptomatic infection and suggest that a major adaptation of the ehrlichiae is the ability to amplify through the mammalian host that remains clinically well. The obvious major advantage to the ehrlichia is the ability to obtain horizontal transmission among tick vectors. Likewise, the ability to alter host immune and defense functions would be another survival advantage to ensure adequate bacteremia for horizontal tick transmission. Whether such mechanisms for persistent infection leading to symptomatic illness occur in humans is still not known. Such critical questions require serious consideration and may explain some of the serious and prolonged manifestations of human illness after tick bite.

10. SUMMARY

Ehrlichiosis caused by *E. phagocytophila*-group ehrlichiae is a zoonotic infection transmitted by tick bite. Humans who are accidentally bitten by *Ixodes* ticks that harbor these obligate intracellular bacteria between stages may develop mild to severe or fatal infections. Laboratory evidence of acquired immunity develops in most patients normally and may persist for years. However, some patients may have active infection even in the presence of serum antibody and severely affected individuals may develop secondary opportunistic infections that lead to life-threatening illness. Although the specific abnormalities in host immune and defense function that allow these superinfections of humans are not known, animal models predict that acute *E. phagocytophila*-group ehrlichia infection may allow a multifactorial defect of cell-mediated immunity and nonspecific phagocytes and inflammation. The critical interaction of persistence and immune and host defense dysfunction may be appropriate for reservoir animals that develop little or no discernible clinical illness, but may allow for severe human disease caused by opportunistic pathogens that are acquired incidentally or concurrently by the bite of a single tick.

REFERENCES

1. Rikihisa, Y., 1991, The tribe *Ehrlichieae* and ehrlichial diseases, *Clin. Microbiol. Rev.* **4:**286–308.
2. Anderson, B. E., Dawson, J. E., Jones, D. C., and Wilson, K. H., 1991, *Ehrlichia chaffeensis*, a new species associated with human ehrlichiosis, *J. Clin. Microbiol.* **29:**2838–2842.
3. Dumler, J. S., and Bakken, J. S., 1995, Ehrlichial diseases of humans: Emerging tick-borne infections, *Clin. Infect. Dis.* **20:**1102–1110.
4. Messick, J. B., and Rikihisa, Y., 1993, Characterization of *Ehrlichia risticii* binding, internalization, and proliferation in host cells by flow cytometry, *Infect. Immun.* **61:**3803–3810.
5. Wells, M., and Rikihisa, Y., 1988, Lack of lysosomal fusion with phagosomes containing *Ehrlichia risticii* in P388D1 cells: Abrogation of inhibition with oxytetracycline, *Infect. Immun.* **56:**3209–3215.
6. Park, J., and Rikihisa, Y., 1991, Inhibition of *Ehrlichia risticii* infection in murine peritoneal macrophages by gamma interferon, a calcium ionophore, and concanavalin A, *Infect. Immun.* **59:**3418–3423.
7. Walker, D. H., and Dumler, J. S., 1996, Emergence of ehrlichioses as human health problems, *Emerg. Infect. Dis.* **2:**18–29.
8. Chen, S.-M., Dumler, J. S., Bakken, J. S., and Walker, D. H., 1994, Identification of a granulocytotrophic *Ehrlichia* species as the etiologic agent of human disease, *J. Clin. Microbiol.* **32:**589–595.
9. Dumler, J. S., Asanovich, K. M., Bakken, J. S., Richter, P., Kimsey, R., and Madigan, J. E., 1995, Serologic cross-reaction among *Ehrlichia equi*, *Ehrlichia phagocytophila*, and human granulocytic ehrlichia, *J. Clin. Microbiol.* **33:**1098–1103.
10. Madigan, J. E., Richter, P. J., Kimsey, R. B., Barlough, J. E., Bakken, J. S., and Dumler, J. S., 1995, Transmission and passage in horses of the agent of human granulocytic ehrlichiosis, *J. Infect. Dis.* **172:**1141–1144.
11. Barlough, J. E., Madigan, J. E., DeRock, E., Dumler, J. S., and Bakken, J. S., 1995, Protection against *Ehrlichia equi* is conferred by prior infection with the human granulocytotropic ehrlichia (HGE agent), *J. Clin. Microbiol.* **33:**3333–3334.
12. Goodman, J. L., Nelson, C., Vitale, B., Dumler, J. S., Madigan, J. E., Kurtti, T. J., and Munderloh, U. G., 1996, Direct cultivation of the causative agent of human granulocytic ehrlichiosis, *N. Engl. J. Med.* **334:**262–263.
13. Munderloh, U. G., Madigan, J. E., Dumler, J. S., Goodman, J. L., Hayes, S. F., Barlough, J. E., Nelson, C. M., and Kurtti, T. J., 1996, Isolation of the equine granulocytic erlichiosis agent, *Ehrlichia equi*, in tick cell culture, *J. Clin. Microbiol.* **34:**664–670.

14. Dumler, J. S., Asanovich, K. M., Madigan, J. E., Barlough, J. E., Bakken, J. S., Aguero-Rosenfeld, M., Wormser, G., and Goodman, J. L., 1996, Diagnosis of granulocytic ehrlichiosis by cell culture, in: *Abstracts of the 96th General Meeting of the American Society for Microbiology, New Orleans* abstr. D-20.

15. Sumner, J. W., Nicholson, W. L., Childs, J. E., and Massung, R. F., 1996, PCR amplification and comparison of nucleotide sequences from the groE heat-shock operon of *Ehrlichia* spp, in: *Abstracts of the Twelfth Sesqui-Annual Meeting of the American Society for Rickettsiology and Rickettsial Diseases, Pacific Grove, CA* abstr. 22.

16. Dawson, J. E., Anderson, B. E., Fishbein, D. B., Sanchez, J. L., Goldsmith, C. S., Wilson, K. H., and Duntley, C. W., 1991, Isolation and characterization of an *Ehrlichia* sp. from a patient diagnosed with human ehrlichiosis, *J. Clin. Microbiol.* **29:**2741–2745.

17. Bakken, J. S., Dumler, J. S., Chen, S. M., Eckman, M. R., Van Etta, L. L., and Walker, D. H., 1994, Human granulocytic ehrlichiosis in the upper midwest United States. A new species emerging? *Am. Med. Assoc.* **272:**212–218.

18. Greig, B., Asanovich, K. M., Armstrong, J. P., and Dumler, J. S., 1996, Geographic, clinical, serologic, and molecular evidence of granulocytic ehrlichiosis, a likely zoonotic disease in Minnesota and Wisconsin dogs, *J. Clin. Microbiol.* **34:**44–48.

19. Madigan, J. E., Barlough, J. E., Dumler, J. S., Schankman, N. S., and DeRock, E., 1996, Equine granulocytic ehrlichiosis in Connecticut caused by an agent resembling the human granulocytic ehrlichia, *J. Clin. Microbiol.* **34:**434–435.

20. Madigan, J. E., and Gribble, D., 1987, Equine ehrlichiosis in northern California: 49 cases (1968–1981), *J. Am. Vet. Med. Assoc.* **190:**445–448.

21. Johansson, K.-E., Pettersson, B., Uhlén, M., Gunnarsson, A., Malmqvist, M., and Olsson, E., 1995, Identification of the causative agent of granulocytic ehrlichiosis in Swedish dogs and horses by direct solid phase sequencing of PCR products from the 16S rRNA gene, *Res. Vet. Sci.* **58:**109–112.

22. Magnarelli, L. A., Dumler, J. S., Anderson, J. F., Johnson, R. C., and Fikrig, E., 1995, Coexistence of antibodies to tick-borne pathogens of babesiosis, ehrlichiosis, and Lyme borreliosis in human sera, *J. Clin. Microbiol.* **33:**3054–3057.

23. Pancholi, P., Kolbert, C. P., Mitchell, P. D., Reed, K. D., Dumler, J. S., Bakken, J. S., Telford, S. R., III, and Persing, D. H., 1995, *Ixodes dammini* as a potential vector of human granulocytic ehrlichiosis, *J. Infect. Dis.* **172:**1007–1012.

24. Mitchell, P. D., Reed, K. D., and Hofkes, J. M., 1996, Immunoserologic evidence of coinfection with *Borrelia burgdorferi, Babesia microti*, and human granulocytic *Ehrlichia* species in residents of Wisconsin and Minnesota, *J. Clin. Microbiol.* **34:**724–727.

25. Brouqui, P., Dumler, J. S., Lienhard, R., Brossard, M., and Raoult, D., 1995, Human granulocytic ehrlichiosis in Europe, *Lancet* **346:**782–783.

26. Bakken, J. S., Krueth, J., Wilson-Nordskog, C., Tilden, R. L., Asanovich, K., and Dumler, J. S., 1996, Clinical and laboratory characteristics of human granulocytic ehrlichiosis, *Am. Med. Assoc.* **275:**199–205.

27. Edelman, D. C., and Dumler, J. S., 1996, Evaluation of an improved PCR diagnostic assay for human granulocytic ehrlichiosis, *Mol. Diagn.* **1:**41–49.

28. Telford, S. R., III, Lepore, T. H., Snow, P., and Dawson, J. E., 1995, Human granulocytic ehrlichiosis in Massachusetts, *Ann. Intern. Med.* **123:**277–279.

29. Hardalo, C., Quagliarello, V., and Dumler, J. S., 1995, Human granulocytic ehrlichiosis in Connecticut: Report of a fatal case, *Clin. Infect. Dis.* **21:**910–914.

30. Hudson, J. R., 1950, The recognition of tick-borne fever as a disease of cattle, *Br. Vet. J.* **106:**3–17.

31. Gordon, W. S., Brownlee, A., Wilson, D. R., and MacLeod, J., 1932, Tick-borne fever. (A hitherto undescribed disease of sheep.) *J. Comp. Pathol. Ther.* **65:**301–307.

32. Foggie, A., 1951, Studies on the infectious agent of tick-borne fever in sheep, *J. Pathol. Bacteriol.* **63**:1–15.
33. Tuomi, J., and von Bonsdorff, C. H., 1996, Electron microscopy of tick-borne fever agent in bovine and ovine phagocytizing leukocytes, *J. Bacteriol.* **92**:1478–1492.
34. MacLeod, J. R., and Gordon, W. S., 1933, Studies in tick-borne fever of sheep. I. Transmission by the tick, *Ixodes ricinus*, with a description of the disease produced, *Parasitology* **25**:273–285.
35. Foggie, A., 1956, The effect of tick-borne fever on the resistance of lambs to staphylococci, *J. Comp. Pathol.* **66**:278–285.
36. Taylor, A. W., Holman, H. H., and Gordon, W. S., 1941, Attempts to reproduce the pyaemia associated with tick-bite, *Vet. Rec.* **53**:337–344.
37. Foster, W. N. M., and Cameron, A. E., 1968, Thrombocytopenia in sheep associated with experimental tick-borne infection, *J. Comp. Pathol.* **78**:251–254.
38. Gordon, W. S., Brownlee, W. S., Brownlee, A., Wilson, D. R., and MacLeod, J., 1932, Studies in louping-ill, *J. Comp. Pathol. Ther.* **45**:106–140.
39. Munro, R., Hunter, A. R., MacKenzie, G., and McMartin, D. A., 1982, Pulmonary lesions in sheep following experimental infection by *Ehrlichia phagocytophila* and *Chlamydia psittaci*, *J. Comp. Pathol.* **92**:117–129.
40. Foster, W. N. M., Foggie, A., and Nisbet, D. I., 1968, Haemorrhagic enteritis in sheep experimentally infected with tick-borne fever, *J. Comp. Pathol.* **78**:255–258.
41. Brodie, T. A., Holmes, P. H., and Urquhart, G. M., 1986, Some aspects of tick-borne diseases of British sheep, *Vet. Rec.* **118**:415–418.
42. Foggie, A., 1957, Further experiments on the effect of tick-borne fever infection on the susceptibility of lambs to staphylococci, *J. Comp. Pathol.* **67**:369–377.
43. Webster, K. A., and Mitchell, G. B. B., 1986, Experimental production of tick pyaemia, *Vet. Rec.* **119**:186–187.
44. Gilmour, N. J. L., Brodie, T. A., and Holmes, P. H., 1982, Tick-borne fever and pasteurellosis in sheep, *Vet. Rec.* **111**:512.
45. Øveras, J., 1972, Disease in sheep kept on *Ixodes ricinus* infected pastures, *Nor. Vet. Tidsskr.* **83**:561–567.
46. Grönestol, H., and Ulvund, M. J., 1977, Listerial septicaemia in sheep associated with tick borne fever (*Ehrlichia ovis*), *Acta Vet. Scand.* **18**:575–577.
47. Batungbacal, M. R., and Scott, G. R., 1982, Tick-borne fever and concurrent parainfluenza-3 virus infection in sheep, *J. Comp. Pathol.* **92**:415–428.
48. Gribble, D. H., 1969, Equine ehrlichiosis, *J. Am. Vet. Med. Assoc.* **155**:462–469.
49. Stannard, A. A., Gribble, D. H., and Smith, R. S., 1969, Equine ehrlichiosis: A disease with similarities to tick-borne fever and bovine petechial fever, *Vet. Rec.* **84**:149–150.
50. Madigan, J. E., Hietala, S., Chalmers, S., and DeRock, E., 1990, Seroepidemiologic survey of antibodies to *Ehrlichia equi* in horses of northern California, *J. Am. Vet. Med. Assoc.* **196**:1962–1964.
51. Brewer, B. D., Harvey, J. W., Mayhew, I. G., and Simpson, C. F., 1984, Ehrlichiosis in a Florida horse, *J. Am. Vet. Med. Assoc.* **185**:446–447.
52. Berrington, A., Moats, R., and Lester, S., 1996, A case of *Ehrlichia equi* in an adult horse in British Columbia, *Can. Vet. J.* **37**:174–175.
53. Bjoersdorff, A., Johnsson, A., Sjöstrom, A. C., and Madigan, J. E., 1990, *Ehrlichia equi*-infektion diagnostiserat hos hast, *Sven. Vet. Tidn.* **42**:357–360.
54. Korbutiak, E., and Schneiders, D., 1994, Equine granulocytic ehrlichiosis in the UK, *Vet. Rec.* **135**:387.
55. McNamee, P. T., Cule, A. P., and Donnelly, J., 1989, Suspected ehrlichiosis in a gelding in Wales, *Vet. Rec.* **124**:634–635.
56. de Alvarado, C. M. A., Finol, G., Parra, O., Riquelme, M., and Savedra, A., 1992, Equine ehrlichiosis in the Zulia State, Venezuela. Report of 232 cases. *Rev. Cient. F.C.V. de Luz* **2**:41–52.

57. Sells, D. M., Hildebrandt, P. K., Lewis, G. E., Nyindo, M. B. A., and Ristic, M., 1976, Ultra-structural observations on *Ehrlichia equi* organisms in equine granulocytes, *Infect. Immun.* **13:**273–280.

58. Richter, P. J., Kimsey, R. B., Madigan, J. E., Barlough, J. E., Brooks, D. L., and Dumler, J. S., 1996, *Ixodes pacificus* as a vector of *Ehrlichia equi, J. Med. Entomol.* **33:**1–5.

59. Madewell, B. R., and Gribble, D. H., 1982, Infection in two dogs with an agent resembling *Ehrlichia equi, J. Am. Vet. Med. Assoc.* **180:**512–514.

60. Rodgers, S. J., Morton, R. J., and Baldwin, C. A., 1989, A serological survey of *Ehrlichia canis, Ehrlichia equi, Rickettsia rickettsii,* and *Borrelia burgdorferi* in dogs in Oklahoma, *J. Vet. Diagn. Invest.* **1:**154–159.

61. Maretzki, C. H., Fisher, D. J., and Greene, C. E., 1994, Granulocytic ehrlichiosis and meningitis in a dog, *J. Am. Vet. Med. Assoc.* **205:**1554–1556.

62. Foster, W. N. M., and Cameron, A. E., 1970, Observations on ovine strains of tick-borne fever, *J. Comp. Pathol.* **80:**429–436.

63. Foggie, A., and Hood, C. S., 1961, Adaptation of the infectious agent of tick-borne fever to guinea-pigs and mice, *J. Comp. Pathol.* **71:**414–427.

64. Lewis, G. E., Huxsoll, D. L., Ristic, M., and Johnson, A. J., 1975, Experimentally induced infection of dogs, cats, and nonhuman primates with *Ehrlichia equi,* etiologic agent of equine ehrlichiosis, *Am. J. Vet. Res.* **36:**85–88.

65. Dumler, J. S., Asanovich, K. M., Bakken, J. S., Madigan, J. E., and Greig, B., 1995, Specific 16S rDNA nucleotide sequences correspond to geographic variants of granulocytic ehrlichiae, *FASEB J.* **9:**A273.

66. Batungbacal, M. R., Scott, G. R., and Burrells, C., 1982, The lymphocytopaenia in tick-borne fever, *J. Comp. Pathol.* **92:**403–407.

67. Woldehiwet, Z., 1987, Depression of lymphocyte response to mitogens in sheep infected with tick-borne fever, *J. Comp. Pathol.* **97:**637–643.

68. Larsen, H. J. S., Øvernes, G., Waldeland, H., and Johansen, G. M., 1994, Immunosuppression in sheep experimentally infected with *Ehrlichia phagocytophila, Res. Vet. Sci.* **56:**216–224.

69. Batungbacal, M. R., and Scott, G. R., 1982, Suppression of the immune response to clostridial vaccine by tick-borne fever, *J. Comp. Pathol.* **92:**409–413.

70. Woldehiwet, Z., 1991, Lymphocyte subpopulations in peripheral blood of sheep experimentally infected with tick-borne fever, *Res. Vet. Sci.* **51:**40–43.

71. Foster, W. N. M., and Cameron, A. E., 1970, Observations on the functional integrity of neutrophil leucocytes infected with tick-borne fever, *J. Comp. Pathol.* **80:**487–491.

72. Woldehiwet, Z., 1987, The effects of tick-borne fever on some functions of polymorphonuclear cells of sheep, *J. Comp. Pathol.* **97:**481–485.

73. Dumler, J. S., and Bakken, J. S., 1996, Human granulocytic ehrlichiosis in Wisconsin and Minnesota: A frequent infection with the potential for persistence, *J. Infect. Dis.* **173:**1027–1030.

74. Woldehiwet, Z., and Scott, G. R., 1982, Immunological studies on tick-borne fever in sheep, *J. Comp. Pathol.* **91:**457–467.

75. Stuen, S., Hardeng, F., and Larsen, H. J., 1992, Resistance to tick-borne fever in young lambs, *Res. Vet. Sci.* **52:**211–216.

76. Nyindo, M. B. A., Ristic, M., Lewis, G. E., Jr., Huxsoll, D. L., and Stephenson, E. H., 1978, Immune response of ponies to experimental infection with *Ehrlichia equi, Am. J. Vet. Res.* **39:**15–18.

77. Tyzzer, E. E., 1938, *Cytoecetes microti* n. gen. n. sp. A parasite developing in granulocytes and infection in small rodents, *Parasitology* **30:**242–257.

10

The Immune Response to *Ehrlichia chaffeensis*

PHILIPPE BROUQUI and J. STEPHEN DUMLER

1. INTRODUCTION

Ehrlichia chaffeensis is the agent of human monocytic ehrlichiosis. The first diagnosed case of disease occurred in a 51-year-old man who became ill in April 1986, 12–14 days after tick bites in rural Arkansas.[1] His severe course of illness was characterized by fever, hypotension, confusion, acute renal failure requiring hemodialysis, pancytopenia, coagulopathy, cutaneous and gastrointestinal hemorrhages, and hepatocellular injury. The diagnosis of ehrlichial infection was documented by the observation of 2- to 5-µm morulae in 1–2% of circulating mononuclear cells. Electron microscopy demonstrated that the inclusions represented membrane-bound vacuoles containing up to 40 bacteria each with a diameter of 0.2–0.8 µm and a gram-negative-type cell wall. At that time the infectious etiology was attributed to *Ehrlichia canis*, a pathogen of dogs, because of a serologic reaction to that organism. *E. chaffeensis,* the etiologic agent of human monocytic ehrlichiosis, was later isolated from the blood of a patient at Ft. Chaffee, Arkansas, in July 1990, who had fever, headache, pharyngitis, nausea, vomiting, and dehydration.[2]

PHILIPPE BROUQUI • Unité des Rickettsies, CNRS-UPRES-A, Faculté de Médecine, 13385 Marseille cedex 5, France.   J. STEPHEN DUMLER • Division of Medical Microbiology, Department of Pathology, The Johns Hopkins Medical Institutions, Baltimore, Maryland 21287-7093.

Rickettsial Infection and Immunity, edited by Anderson *et al.* Plenum Press, New York, 1997.

2. EPIDEMIOLOGY

E. chaffeensis has been detected in two tick species, *Amblyomma americanum* (the lone star tick)[3] and *Dermacentor variabilis* (the American dog tick).[4] White-tailed deer of most endemic areas have antibodies reactive with *E. chaffeensis*.[5] These deer are susceptible to experimental infection with ehrlichiae that can circulate for weeks.[5] It seems likely that deer or other animals such as dogs or small rodents serve as reservoir host for *E. chaffeensis*, and that *A. americanum* ticks serve as the major vector. Indeed, under experimental conditions, larval and nymphal *A. americanum* acquire *E. chaffeensis* by feeding on infected deer, maintain the ehrlichiae transstadially, and transmit *E. chaffeensis* while feeding as nymphs and adults on naive deer.[6]

More than 400 cases of human monocytic ehrlichiosis have been reported in 30 states[7] of the United States. That these most likely represent just the tip of the iceberg was confirmed when MRL Diagnostics, a commercial reference laboratory that offers IFA serology for *E. chaffeensis*, reported 722 positive specimens between September 1992 and June 15, 1995. The infection in humans is considered reportable in only a few states. Because relatively few physicians even know that these diseases exist, most infections are not diagnosed. Even highly knowledgeable physicians find ehrlichiosis difficult to diagnose on the basis of clinical signs and symptoms. Most patients (83%) report exposure to ticks or tick bite within 3 weeks of onset of illness.[8,9] Cases are predominantly rural (66%) and seasonal (68% May–July). The median age of patients is 44 years, and three-quarters are male. Outbreaks of human monocytic ehrlichiosis among groups of golfers emphasize the risk for infection during outdoor activities with tick exposure.[10,11] Human monocytic ehrlichiosis occurs not only as an acute illness or prolonged fever of undetermined origin[12] in apparently immunocompetent persons but also as an opportunistic infection of patients with compromised host defenses, including AIDS patients or steroid-treated patients.[12–16] The report of fatal cases in HIV-infected patients, in whom diagnostic antibody responses to *E. chaffeensis* never developed, suggests that such cases would not usually be diagnosed correctly.

3. PATHOGENESIS

E. chaffeensis is introduced into the dermis by the bite of an infected tick and spreads hematogenously throughout the body. Intracellular infection is established within phagosomes, most often in macrophages in the liver, spleen, and lymph nodes, but other sites, including bone marrow, lung, kidney, and CSF, can be involved.[17–20] Pathologic lesions associated with ehrlichial infection include focal necroses of the liver, spleen, and lymph nodes, multiorgan perivascular lymphohistiocytic infiltrates, hemophagocytosis in the spleen, liver, bone marrow, and lymph nodes, interstitial pneumonitis, and pulmonary hemorrhage. The most striking discovery in bone marrow was the frequent occurrence of granulomas as manifestations of cell-mediated immune response to this organism.

The pathogenic mechanisms of ehrlichial disease are poorly understood. *E. chaffeensis* directly causes necrosis of heavily infected cells *in vitro* and in immunocompromised patients;[12] however, the role of host immune and inflammatory responses as disease mechanisms has yet to be determined. Observation of opportunistic fungal and viral infections in severe and fatal cases suggests the possibility of an ehrlichial role in the suppression or dysregulation of the immune response.[1,17]

4. VIRULENCE OF *E. CHAFFEENSIS*

Currently only two isolates of *E. chaffeensis* have been reported cultivated from humans.[2,21] The Arkansas strain was isolated from a patient with a mild disease[2] and the 91HE17 strain was isolated from a nearly fatal case of human ehrlichiosis with respiratory distress, acute renal failure, and meningitis.[21] Minor molecular differences observed in the 16 S rRNA gene nucleotide sequences of these isolates may reflect other differences in the bacterial genome that encode variant proteins associated with differences in virulence. The existence of potential variances is supported by data that show reaction of the monoclonal antibody 6A1 directed against a 30-kDa protein antigen of *E. chaffeensis* Arkansas strain but not with proteins of the 91HE17 strain. With proteins that range from 22 to 30 kDa having been suggested as species-specific antigens,[22] it is likely that these are candidate molecules for involvement in the expression of disease in humans. Nevertheless, such a hypothesis needs to be demonstrated by cultivation and further analysis of other isolates of *E. chaffeensis* from humans.

5. IMMUNITY TO *E. CHAFFEENSIS*

Little is known of specific immunity to *E. chaffeensis* partly because of the only recent availability of the bacteria in cell culture. However, other closely related ehrlichiae have been studied, especially *E. risticii, E. sennetsu, E. canis,* and *Cowria ruminantium,* all monocytic ehrlichiae. Although ehrlichiae are thought to be highly host specific, the immunologic responses among these monocytic ehrlichiae are very similar and one may consider that immunity to *E. chaffeensis* is in some manner also very similar.

TABLE I
Clinical Presentation and Host Factors in Severe Human Monocytic
Ehrlichiosis (*Ehrlichia chaffeensis* Infection)

Clinical presentation	Age	Host factors	Severity	IFA titers	Ref.
1. Fever, pneumonia, pancytopenia, vaginal bleeding	41	HIV+	Fatal	<64	12
2. Fever, pneumonia, rash, gastrointestinal bleeding	65	?	Fatal	1,280	18
3. Fever, pancytopenia, DIC, septic shock	67	?	Recovered	1,280	24
4. Fever, ARDS, acute renal failure, lymphocytic meningitis	72	?	Recovered	32,768	19
5. Respiratory failure, seizure and coma, acute renal failure, severe cholestasis	56	Sulfadiazine + ulcerative colitis	Recovered	5,120	28
6. Fever, pneumonia, myocarditis	43	?	Recovered	1,024	27
7. Fever, scleral icterus	51	Liver transplant + cyclosporine + prednisone	Recovered	2,560	26
8. Fever, confusion, lymphocytic meningitis, severe pancytopenia, respiratory failure	68	Lobectomy lung abscess + alcohol intake	Fatal	2,560	17
9. ARDS, encephalitis, pancreatitis, toxic shock syndrome	46	Splenectomy	Recovered	256	13
10. Toxic shock syndrome-like	37	HIV+	Recovered	1,024	13
11. Fever, ARDS, acute renal failure, lymphocytic meningitis	51	?	Recovered	640	1
12. Pneumonia, meningitis	62	?	Recovered	640	25
13. Fever, meningismus, renal failure, severe thrombocytopenia	67	High-dose corticosteroid therapy	Fatal	512	15
14. Fever, pancytopenia, renal failure	36	HIV+	Fatal	n.d.[a]	16

[a]n.d., not done.

5.1. The Role of Host Factors in the Clinical Presentation

The clinical course of illness ranges from asymptomatic to fatal. The two patients from whom *E. chaffeensis* has been isolated reflect this variation in clinical severity. Differences in virulence between isolates of *E. chaffeensis* may account for differences in severity; however, it is equally likely that underlying host illnesses or other host factors may be determinants of severity as well. Fatal outcome occurs in 2% of infected patients.[23] Death occurs mostly in patients age 60 and older and in immunocompromised patients (Table I). In reported severe cases of documented human monocytic ehrlichiosis, many patients were children or were older than 60 years (case Nos. 2, 3, 4, 8, 12, and 13)[18,19,21,24,25] or had a known T-cell abnormality such as HIV infection (case Nos. 1, 10, and 14),[12,13,16] organ transplant (case No. 7),[14,26] corticosteroid therapy (case No. 13), or were splenectomized (case No. 9).[13] Only 3 patients did not have a known underlying condition.[1,27,28] In a logistic regression analysis of 237 patients whose serum had a fourfold increase or decrease in antibody titer to *E. chaffeensis*, it was found that the probability of severe illness or death was higher among case-patients 60 years or older.[9] Severity of rickettsioses, especially Rocky Mountain spotted fever and Mediterranean spotted fever, has been reported to be associated with age, alcohol abuse, T cell depression, or G-6-PD deficiency.[29] It has been suggested that iron overload could play a role in the severity of rickettsioses as is true for other intracellular infections such as yersiniosis.[29]

5.2. The Role of Macrophages in Intracellular Killing of Ehrlichiae

Killing of intracellular bacteria by monocytes includes oxygen-dependent and -independent mechanisms. The oxidative burst seems to play a minor role in killing of intracellular ehrlichiae. Peritoneal macrophages from two strains of mice, BALB/c and C3H/HeN, failed to respond with a respiratory burst after phagocytosis of *E. risticii*, and bactericidal effects of concanavalin A and the calcium ionophore A23187 on *E. risticii* are related to an IFN-γ-like effect rather than to induction of an oxidative metabolic burst.[30,31] It is likely that the mechanism by which ehrlichiae survive and multiply in the infected cell relies on the ability to inhibit phagosome–lysosome fusion. This has been clearly demonstrated for *E. risticii* and *E. sennetsu*.[32,33] As phagosome–lysosome fusion is restored by subinhibitory concentrations of doxycyline (an antibiotic that inhibits prokaryote protein synthesis) but not by other effective antibiotics, it has been suggested that phagosome–lysosome fusion inhibition is related to a secreted bacterial protein.[33] Thus, it is likely that T-cell induced immunity and IFN-γ secretion is the predominant mechanism for recovery from and immunity to ehrlichia infection. In fact, this is generally true for most intracellular bacteria, as humoral or antibody-mediated immunity acts predominantly by enhancing opsonization and phagocytosis.

The role of IFN-γ and other cytokines on subsequent killing of intracellular ehrlichiae has been studied for *E. risticii* and *E. chaffeensis*.[34–36] Several facultative intracellular pathogens which are known to survive and multiply in the membrane-lined vacuoles of macrophages are either killed or stop proliferating *in vitro* when macrophages are treated with IFN-γ prior to or after infection with the pathogen. These mechanisms are associated with an increase in phagosome–lysosome fusion, downregulation of the transferrin receptor and subsequent intracellular iron depletion, indoleamine 2,3-dioxygenase activation, or nitrate production presumably secondary to induction of nitric oxide synthetase.[36] *In vitro* treatment of mouse peritoneal macrophages with recombinant murine IFN-γ confers resistance to infection with *E. risticii*, and allows eradication of ehrlichiae in infected macrophages.[36] It is important that IFN-γ added after infection of macrophages had an ehrlichiacidal effect as this more closely mimics the natural time sequence. Thus, IFN-γ most likely plays a key role in the elimination of established ehrlichia. IFN-γ-pretreated human blood monocytes and human monocyte cell line THP-1 were able to inhibit *E. chaffeensis* growth *in vitro* when added before or within 6 hr after infection.[34] The ehrlichiacidal activity of IFN-γ-treated peritoneal macrophages is suppressed by N^G-monomethyl-*l*-arginine, a competitive inhibitor of nitric oxide synthesis from *l*-arginine.[36] Moreover, the intracellular iron chelator, deferoxamine, also inhibited *E. risticii* infection *in vitro*. This indicates that the generation of nitric oxide accompanied by limitation in access to iron is the mechanism by which IFN-γ inhibits *E. risticii* growth *in vitro*. Nitric oxide production was not demonstrated in the medium of IFN-γ-treated *E. chaffeensis*-infected monocytes, whereas, as shown with *E. risticii* infection, deferoxamine completely inhibits the survival of the bacteria. Addition of iron-saturated transferrin to cultures was able to reverse the IFN-γ-induced ehrlichial killing in a dose-dependent manner. This indicates that IFN-γ-activated monocytes inhibit infection by *E. chaffeensis* by downregulation of surface transferrin receptors that in turn leads to limitation of available cytoplasmic iron rather than by nitric oxide generation as reported for *E. risticii*.[34] The divergent findings may reflect the different microbicidal mechanisms activated in each cell culture–infection system or differences in the sensitivity of each organism to the variety of microbicidal actions of monocyte / macrophage cells.

5.3. The Role of Specific Immune Response to *E. chaffeensis*

Antibody-Mediated Immunity

Serological cross-reactions are common findings among some species of the genus *Ehrlichia*. This explains the first reported case of human ehrlichiosis in the United States that was attributed to *E. canis* instead of *E. chaffeensis*. This cross-reaction was further used to screen sera from patients with ehrlichiosis until *E. chaffeensis* became available as a source of antigen. Western blot analysis of sera from patients with human ehrlichiosis and from dogs with *E. canis* infection showed that 27- to 29-kDa proteins were revealed only by *E. chaffeensis* antisera whereas a 22- to 25-kDa protein was specifically detected by anti-*E. canis* serum.[22,37,38] Dogs infected with *E. chaffeensis* reacted with the 29-kDa protein of *E. chaffeensis* but poorly with *E. canis* antigen. Similarly, antisera of *E. canis*-infected dogs reacted strongly with the 30-kDa protein of *E. canis* and poorly with *E. chaffeensis* antigen.[37] Thus, both human infection with *E. chaffeensis* and the canine disease due to *E. canis* cause specific antibody responses that can be distinguished by Western immunoblotting.

These humoral immune responses correlate with results of animal experiments showing that *E. chaffeensis*-infected pups that were subsequently challenged with *E. canis* became severely ill indicating the lack of cross protection between those two organisms[39] as has been previously reported between two other serologically similar ehrlichiae, *E. sennetsu* and *E. risticii*.[40] Thus, humoral immune responses to shared antigens do not play a major role in immunity to ehrlichiae. It is likely that antibodies enhance phagocytosis by opsonization as demonstrated when the closely related *E. canis*, which grows in canine macrophages in the presence of normal canine serum and causes macrophage death, are killed by canine macrophages when the ehrlichiae are cultivated in medium supplemented with immune serum.[41,42]

Antigen presentation is a key component in induction of immunity toward intracellular bacteria. Indirect immunofluorescence staining of *E. risticii*-infected macrophages (P388D$_1$) with an anti-*E. risticii* serum revealed a punctate staining pattern on the surface of the host cell that differed from that of bound bacteria. As the intracellular ehrlichial burden increased, the amount of ehrlichial antigen on the surface increased. This increase was inhibited by tetracycline treatment of infected cells as well as by pronase treatment suggesting that the surface-presented antigen was of ehrlichial protein origin.[26] In fact, antibody-dependent cellular cytotoxicity may play a role in the lysis of ehrlichia-infected cells.[26]

Cell-Mediated Immunity

Absolute lymphopenia is one of the most frequent laboratory findings in human monocytic ehrlichiosis.[23] However, after doxycycline therapy lymphocytosis occurs in many cases.[43] The reason for the initial lymphopenia is not known. The lymphocytosis that occurs in human monocytic ehrlichiosis is of γ/δ T-cell type. This is a common finding in other infectious diseases although the percentage of such cells is usually low (<25%) whereas as many as 97% of the lymphocytes have this phenotype after ehrlichiosis.[43]

Human monocytic ehrlichiosis is sometimes severe and investigation is needed to better understand the pathogenesis and the host immune response that aid in recovery from infection. The lack of prior studies was mostly related to the unavailability of a system for continuous tissue culture propagation. As for other intracellular organisms, cell-mediated immunity is likely to be important in ehrlichiosis. Adequate development of T-cell immunity to *E. chaffeensis* infection in humans may explain why this disease is usually not severe. However, increased severity with a potential for death occurs when patients are immunocompromised by conditions such as AIDS, cancer, organ transplantation, and extremes in age. Thus, monocytic ehrlichiosis should be systematically considered in immunocompromised patients and/or with fever of unknown origin, especially if the patient has had a tick bite or if the patient lives in an endemic area. Specific doxycycline therapy should be initiated as rapidly as possible[44] followed by specific laboratory diagnosis.

REFERENCES

1. Maeda, K., Markowitz, N., Hawley, R. C., Ristic, M., Cox, D., and McDade, J., 1987, Human infection with *Ehrlichia canis*, a leukocytic rickettsia, *N. Engl. J. Med.* **316**:853–856.
2. Dawson, J. E., Anderson, B. E., Fishbein, D. B., Sanchez, J. L., Goldsmith, C. S., Wilson, K. H., and Duntley, C. W., 1991, Isolation and characterization of an *Ehrlichia* from a patient diagnosed with human ehrlichiosis, *J. Clin. Microbiol.* **29**:2741–2745.
3. Anderson, B. E., Sims, K. G., Olson, J. G., Childs, J. E., Piesman, J. F., Happ, C. M., Maupin, G. O., and Johnson, B. J. B., 1993, *Amblyomma americanum:* A potential vector of human ehrlichiosis, *Am. J. Trop. Med. Hyg.* **49**:239–244.
4. Anderson, B. E., Sumner, J. W., Dawson, J. E., Tzianabos, T., Greene, C. R., Olson, J. G., Fishbein, D. B., Olsen-Ramussen, M., Holloway, P. B., Edwin, H. G., and Azad, A. F., 1992, Detection of the etiologic agent of human ehrlichiosis by polymerase chain reaction, *J. Clin. Microbiol.* **30**:775–780.
5. Dawson, J., Stallknech, D. E., Howerth, E. W., Warner, C., Biggie, K., Davidson, W. R., Lockhart, J. M., Nettles, V. F., Olson, J. G., and Childs, J. E., 1994, Susceptibility of white-tailed deer (*Odocoileus virginianus*) to the infection with *Ehrlichia chaffeensis*, the etiologic agent of human ehrlichiosis, *J. Clin. Microbiol.* **32**:2725–2728.
6. Ewing, S. A., Dawson, J. E., Kocan, A. A., Barker, R. W., Warner, C. K., and Panciera, R. J., 1995, Experimental transmission of *Ehrlichia chaffeensis* (Rickettsiales: Ehrlichiae) among white-tailed deer by *Amblyomma americanum* (Acari:Ixodidae), *J. Med. Entomol.* **32**:368–374.
7. Walker, D. H., and Dumler, J. S., 1996, Emergence of the ehrlichioses as human health problems, *Emerg. Infect. Dis.* **2**:18–29.
8. Eng, T. R., Harkess, J. R., Fishbein, D. B., Dawson, J. E., Greene, C. N., Redus, M. A., and Satalowich, F. T., 1990, Epidemiologic, clinical, and laboratory findings of human ehrlichiosis in the United States, *J. Am. Med. Assoc.* **264**:2251–2258.
9. Fishbein, D. B., Dawson, J. E., and Robinson, L. E., 1994, Human ehrlichiosis in the United States, 1985 to 1990, *Ann. Intern. Med.* **120**:736–743.
10. Yevich, S. J., Sanchez, J. L., DeFraites, R. F., Rives, C. C., Dawson, J. E., Uhaa, I. J., Johnson, B. J. B., and Fishbein, D. B., 1995, Seroepidemiology of infections due to spotted fever group rickettsiae and ehrlichia species in military personnel exposed in areas of the United States where such infections are endemic, *J. Infect. Dis.* **175**:1266–1273.
11. Standaert, S. M., Dawson, J. E., Shaffner, W., Childs, J. E., Biggie, K. L., Singleton, J., Jr., Gerhardt, R. R., Knight, M. L., and Hutcheson, R. H., 1995, Ehrlichiosis in a golf-oriented retirement community, *N. Engl. J. Med.* **333**:420–425.
12. Paddock, C. D., Suchard, D. P., Grumbach, K. L., Hadley, W. K. K., Kerschmann, R. L., Abbey, N. W., Dawson, J. E., Anderson, B. E., Sims, K. G., Dumler, J. S., and Herndier, B. G., 1993, Fatal seronegative ehrlichiosis in a patient with HIV infection, *N. Engl. J. Med.* **329**:1164–1167.
13. Fichtenbaum, C. J., Peterson, L. R., and Weil, G. J., 1993, Ehrlichiosis presenting as a life-threatening illness with features of the toxic shock syndrome, *Am. J. Med.* **95**:351–357.
14. Roland, W. E., McDonald, G., Cauldwell, C. W., and Everett, E. D., 1995, Ehrlichiosis—a cause of prolonged fever, *Clin. Infect. Dis.* **20**:821–825.
15. Marty, A. M., Dumler, J. S., Imes, G., Brusman, H. P., Smrkovski, L. L., and Frisman, D. M.,

1995, Ehrlichiosis mimicking thrombotic thrombocytopenic purpura. Case report and pathological correlation, *Hum. Pathol.* **26:**920–925.

16. Barenfanger, J., Patel, P. G., Dumler, J. S., and Walker, D. H., 1996, Human monocytic ehrlichiosis: A new tick-borne infection, *Lab. Med.* **27:**372–374.

17. Dumler, J. S., Sutker, W. L., and Walker, D. H., 1993, Persistent infection with *Ehrlichia chaffeensis*, *Clin. Infect. Dis.* **17:**903–905.

18. Dumler, J. S., Brouqui, P., Aronson, J., Taylor, J. P., and Walker, D. H., 1991, Identification of ehrlichia in human tissue, *N. Engl. J. Med.* **325:**1109–1110.

19. Dunn, B. E., Monson, T. P., Dumler, J. S., Morris, C. C., Westbrook, A. B., Duncan, J. L., Dawson, J. E., Sims, K. G., Anderson, B. E., 1992, Identification of *Ehrlichia chaffeensis* morulae in cerebrospinal fluid mononuclear cells, *J. Clin. Microbiol.* **30:**2207–2210.

20. Dumler, J. S., Dawson, J. E., and Walker, D. H., 1993, Human ehrlichiosis: Hematopathology and immunohistologic detection of *Ehrlichia chaffeensis*, *Hum. Pathol.* **24:**391–396.

21. Dumler, J. S., Chen, S. M., Asanovich, K., Trigiani, E., Popov, V. L., and Walker, D. H., 1995, Isolation and characterization of a new strain of *Ehrlichia chaffeensis* from a patient with nearly fatal monocytic ehrlichiosis, *J. Clin. Microbiol.* **33:**1704–1711.

22. Brouqui, P., Lecam, C., Olson, J. G., and Raoult, D., 1994, Serologic diagnosis of human monocytic ehrlichiosis by immunoblot analysis, *Clin. Diagn. Lab. Immunol.* **1:**645–649.

23. Walker, D. H., and Dumler, J. S., 1995, *Ehrlichia chaffeensis* (human ehrlichiosis) and other ehrlichiae, in: *Principles and Practice of Infectious Diseases, 4th ed.* (G. L. Mandell, J. E. Bennett, and R. Dolin, eds.), Churchill Livingstone, Edinburgh, pp. 1747–1752.

24. Abbott, K. C., Vukelja, S. J., Smith, C. E., McAllister, C. K., Konkol, K. A., O'Rourke, T., Holland, C. J., and Ristic, M., 1991, Hemophagocytic syndrome: A cause of pancytopenia in human ehrlichiosis, *Am. J. Hematol.* **38:**230–234.

25. Dimmitt, D. C., Fishbein, D. B., and Dawson, J. E., 1989, Human ehrlichiosis associated with cerebrospinal fluid pleocytosis, *Am. J. Med.* **87:**677–678.

26. Antony, S. J., Dummer, J. S., and Hunter, E., 1995, Human ehrlichiosis in a liver transplant recipient, *Transplantation* **60:**879–880.

27. Williams, J. D., Snow, R. M., and Arciniegas, J. G., 1995, Myocardial involvement in a patient with human ehrlichiosis, *Am. J. Med.* **98:**414–415.

28. Moskovitz, M., Fadden, R., and Min, T., 1991, Human ehrlichiosis: A rickettsial disease associated with severe cholestasis and multisystemic disease, *J. Clin. Gastroenterol.* **13:**86–90.

29. Walker, D. H., and Raoult, D., 1995, *Rickettsia rickettsii* and other spotted fever group rickettsiae (Rocky Mountain spotted fever and other spotted fevers), in: *Principles and Practice of Infectious Diseases, 4th ed.* (J. L. Mandell, J. E. Bennett, and R. Dolin, eds.), Churchill Livingstone, Edinburgh, pp. 1721–1727.

30. Williams, N. M., Cross, R. J., and Timoney, P. J., 1994, Respiratory burst activity associated with phagocytosis of *Ehrlichia risticii* by mouse peritoneal macrophages, *Res. Vet. Sci.* **57:**194–199.

31. Park, J., and Rikihisa, Y., 1991, Inhibition of *Ehrlichia risticii* infection in murine peritoneal macrophages by gamma interferon, a calcium ionophore, and concanavalin A, *Infect. Immun.* **59:**3418–3423.

32. Wells, M. Y., and Rikihisa, Y., 1988, Lack of lysosomal fusion with phagosomes containing *Ehrlichia risticii* in $P388D_1$ cells: Abrogation of inhibition with oxytetracycline, *Infect. Immun.* **56:**3209–3215.

33. Brouqui, P., and Raoult, D., 1991, Effects of antibiotics on the phagolysosome fusion in *Ehrlichia sennetsu* infected P 388 D1 cells, in: *Rickettsia and Rickettsial Diseases* (J. Kazar and D. Raoult, eds.), Publishing House of the Slovak Academy of Sciences, Bratislava, pp. 751–757.

34. Barnewall, R. E., and Rikihisa, Y., 1994, Abrogation of gamma interferon-induced inhibition of *Ehrlichia chaffeensis* infection in human monocytes with iron transferrin, *Infect. Immun.* **62:**4804–4810.

35. Van Heeckeren, A. M., Rikihisa, Y., Park, J., and Fertel, R., 1993, Tumor necrosis factor alpha, interleukin-1 alpha, interleukin-6 and prostaglandin E2 production in murine peritoneal macrophages infected with *Ehrlichia risticii*, *Infect. Immun.* **61:**4333–4337.

36. Park, J., and Rikihisa, Y., 1992, 1-Arginine-dependent killing of intracellular *Ehrlichia risticii* by macrophages treated with gamma interferon, *Infect. Immun.* **60:**3504–3508.

37. Rikihisa, Y., Ewing, S. A., and Fox, J. C., 1994, Western immunoblot analysis of *Ehrlichia chaffeensis*, *E. canis*, or *E. ewingii* infections in dogs and humans, *J. Clin. Microbiol.* **32:**2107–2112.

38. Chen, S. M., Dumler, J. S., Feng, H. M., and Walker, D. H., 1994, Identification of the antigenic constituents of *Ehrlichia chaffeensis*, *Am. J. Trop. Med. Hyg.* **50:**52–58.

39. Dawson, J. E., and Ewing, S. A., 1992, Susceptibility of dogs to infection with *Ehrlichia chaffeensis* causative agent of human ehrlichiosis, *Am. J. Vet. Res.* **53:**1322–1327.

40. Rikihisa, Y., Pretzman, C. I., Johnson, G. C., Reed, S. M., Yamamoto, S., and Andrews, F., 1988, Clinical, histopathological, and immunological responses of ponies to *Ehrlichia sennetsu* and subsequent *Ehrlichia risticii* challenge, *Infect. Immun.* **56:**2960–2966.

41. Lewis, G. E., Jr., Hill, S. L., and Ristic, M., 1978, Effect of canine immune serum on the growth of *Ehrlichia canis* within non immune canine macrophages, *Am. J. Vet. Res.* **39:**71–76.

42. Lewis, G. E., Jr., and Ristic, M., 1978, Effect of canine immune macrophages and canine immune serum on the growth of *Ehrlichia canis*, *Am. J. Vet. Res.* **39:**77–82.

43. Caldwell, C. W., Everett, E. D., McDonald, G., Yesus, Y. W., and Roland, W. E., 1995, Lymphocytosis of γ/δ T cells in human ehrlichiosis, *Am. J. Clin. Pathol.* **103:**761–766.

44. Brouqui, P., and Raoult, D., 1992, *In vitro* antibiotic susceptibility of the newly recognized agent of ehrlichiosis in humans, *Ehrlichia chaffeensis*, *Antimicrob. Agents Chemother.* **36:**2799–2803.

11

Clinical Aspects of *Bartonella* Infections

LEONARD N. SLATER and DAVID F. WELCH

1. INTRODUCTION

The genus *Bartonella* is named for Dr. A. L. Barton, who described the erythro-cyte-adherent bacterium, *B. bacilliformis*, in 1909. This is the agent of Oroya fever, a syndrome of sepsis and hemolysis, and of verruga peruana, principally a nodular vascular eruption resulting from persistent infection. The link between the two conditions, although previously suspected, was confirmed tragically in 1885 by Daniel Carrión, a medical student who injected himself with bloody material from a verruga and subsequently died of Oroya fever. Consequently, the eponym "Carrión's disease" has been applied to the spectrum of clinical manifestations. Limited to the Andes mountain region of South America, *B. bacilliformis* infection had received little attention outside its endemic zone in recent years until related bacteria, originally classified in the genus *Rochalimaea*, were found to be pathogens in AIDS and other circumstances.

2. TAXONOMY

The former genus *Rochalimaea*, previously grouped with *Bartonella* in the order *Rickettsiales*, had long contained only two member species, *Rochalimaea vinsonii*,

LEONARD N. SLATER • Infectious Diseases Section, The University of Oklahoma Health Sciences Center, and The Department of Veterans Affairs Medical Center, Oklahoma City, Oklahoma 73104.   DAVID F. WELCH • Pediatric Infectious Diseases Section, The University of Oklahoma Health Sciences Center, Oklahoma City, Oklahoma 73104. *Present address of D.F.W.:* Laboratory Corporation of America, Dallas, Texas 75230.

Rickettsial Infection and Immunity, edited by Anderson *et al.* Plenum Press, New York, 1997.

the "Canadian vole agent," and *Rochalimaea quintana*, the agent of trench fever, a debilitating but self-limited human illness so-named after it affected many military personnel in World War I. Except for sporadic outbreaks, trench fever had all but disappeared from the clinical scene in recent decades. However, *R. quintana* reemerged in the 1990s as a pathogen of considerable interest[1-5] coincident with the discovery of two related species pathogenic to humans, originally named *Rochalimaea henselae* and *Rochalimaea elizabethae.*[6-9]

The genus *Bartonella* belongs to the alpha-2 subgroup of the class Proteobacteria, of which *Afipia, Agrobacterium,* and *Brucella* are also members. Sequences of their 16 S ribosomal RNA genes have revealed high levels of relatedness between *B. bacilliformis* and the former *Rochalimaea* spp.[10] and have confirmed that all of them are more closely related to *Brucella* and *Agrobacterium* than to members of the *Rickettsiaceae.* Based on DNA hybridization and 16 S rRNA similarity values, the former *Rochalimaea* species were combined with *Bartonella* in 1993,[11] and the members of the family *Bartonellaceae* were removed from the order *Rickettsiales.* In 1995, an additional merger of a number of species of the genus *Grahamella,* which are intraerythrocytic pathogens of rodents, birds, fish, and other animals, into the genus *Bartonella* was put forth.[12] *Bartonella* spp. can be cultured on cell-free media, unlike members of the order *Rickettsiales.*

3. EPIDEMIOLOGY

B. bacilliformis infections are transmitted naturally only at middle altitudes of the Andes mountains, probably reflecting the distribution of species of the genus *Lutzomyia* (formerly *Phlebotomus*), its sandfly vectors. *Bartonella quintana* is globally distributed, and there have been reports of focal but widely separated outbreaks of trench fever (also known by such eponyms as Volhynia fever, Meuse fever, His-Werner disease, shinbone fever, shank fever, and quintan or five-day fever). Outbreaks often have been associated with conditions of poor sanitation and personal hygiene which may predispose to exposure to the human body louse, the only identified vector of *B. quintana.* Nonhuman vertebrate reservoirs have not yet been identified for either *B. bacilliformis* or *B. quintana.*

B. henselae has been documented to cause bacteremia[13-15] in healthy domestic cats, including some that have been specifically associated with bacillary angiomatosis (BA)[14] or typical cat-scratch disease (CSD)[15] in their human contacts. *B. henselae* could be cultured from the blood of 41% of samples of both pet and impounded cats in San Francisco.[14] A subsequent study from northern California sites found the prevalence of bacteremia to be even higher among impounded (53.0%) or formerly stray cats (70.4%).[16] By contrast, the prevalence of bacteremia among stray cats in Hawaii was found to be notably lower.[15]

Nevertheless, serologic evidence of infection of cats is geographically widespread.[15-20]

Transmission of *B. henselae* to humans has been linked to cats by a number of studies. These studies include serologic and epidemiologic studies[17,21-23] and culture recovery from cases of human lymphadenitis consistent with CSD.[15,24] Identification by polymerase chain reaction (PCR)-based DNA identification in further cases of CSD lymphadenitis[25-27] and conjunctival disease,[28] as well as in CSD skin test antigen[29,30] have further linked *B. henselae*-mediated disease with the domestic cat.

The major suspected arthropod vector of *B. henselae* is the cat flea, based on both epidemiologic associations[15,17,22] and reports of identification of *B. henselae* by both culture and DNA amplification from such fleas.[14,15] It is unclear whether they serve primarily as vectors for cat-to-cat infection or whether they also contribute to human infection.

Like *B. quintana*, *B. henselae* is globally endemic. However, regional variations in the prevalence of either *B. henselae* or *B. quintana* may occur.

4. CLINICAL/PATHOLOGIC MANIFESTATIONS

4.1. Oroya Fever and Verruga Peruana

Oroya fever, the acute illness of *B. bacilliformis* primary bacteremia, usually develops 3 to 12 weeks after inoculation.[31] In its milder, insidiously developing form, the febrile illness often remits in a week. When onset of illness is abrupt, high fever, chills, diaphoresis, headache, and mental status change are associated with a rapidly developing, profound anemia related to bacterial invasion of erythrocytes.[32-34] Intense myalgias and arthralgias, lymphadenopathy, thrombocytopenia, and complications such as seizures, delirium, meningoencephalitis, obtundation, dyspnea, and angina can occur during this stage.[35] Most of these complications are thought to be a consequence of the severe anemia and of microvascular thrombosis, resulting in end-organ ischemia.

Without antimicrobial therapy (chloramphenicol being used most commonly), fatalities can be high.[35] For survivors, convalescence is associated with a decline of fever and disappearance of bacteria on blood smears, but also a temporarily increased susceptibility to subsequent (opportunistic) infections such as salmonellosis[36] or toxoplasmosis.[37,38] Asymptomatic persistent bacteremia with *B. bacilliformis* infection can occur in up to 15% of survivors of acute infection.[39] They may serve as the organism's reservoir.

Verruga peruana usually becomes evident within months of acute infection. This late-stage manifestation is characterized by crops of nodular skin lesions[32] (Fig. 1); mucosal and internal lesions can also occur. Their histology typically

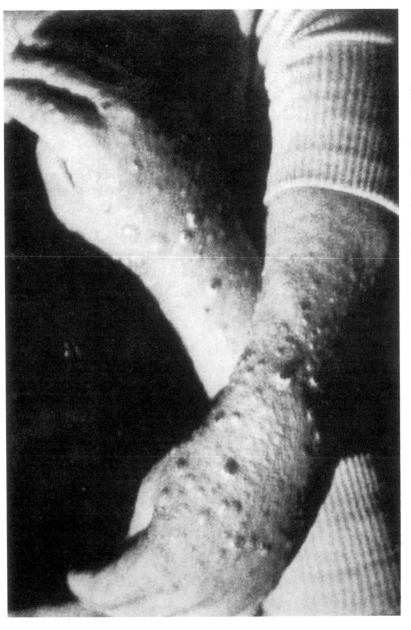

FIGURE 1. Cutaneous lesions of verruga peruana, in a stage of early resolution. (From Strong et al.[32])

contains neovascular proliferation with occasional bacteria evident in interstitial spaces. Bacterial invasion of/replication within endothelial cells (long believed the etiology of cytoplasmic inclusions first described by Rocha-Lima) is rare.[40] The nodules may develop at one site while receding at another. They may persist for months to years, and eventually become fibrotic with involution.

4.2. Bacteremic Illness Related to *B. quintana,* *B. henselae,* Other Species

Acute mortality caused by bacteremia with non-*bacilliformis Bartonella* species, even when persistent, is apparently uncommon. In recent years, *B. quintana* bacteremic infection outside of the context of HIV infection has been identified sporadically, mainly in homeless persons in North America and Europe.[41] Trench fever's natural course includes a spectrum of self-limited clinical patterns.[42,43] Four clinical patterns are recognized. Incubation after inoculation may span 3–38 days before the usually sudden onset of chills and fevers. In the shortest form, a single bout of fever lasts 4–5 days. In the more typical periodic form, there are three to five and sometimes up to eight febrile paroxysms, each lasting about 5 days. The continuous form is manifested by 2–3 weeks up to 6 weeks of uninterrupted fever. Afebrile infection is the least common form. The illness may be accompanied by other nonspecific symptoms and signs such as headache, vertigo, retro-orbital pain, conjunctival injection, nystagmus, myalgias, arthralgias, hepatosplenomegaly, rash, leukocytosis, and albuminuria.

B. quintana or *B. henselae* bacteremia in HIV-infected persons is often characterized by insidious development of symptoms which include malaise, body aches, fatigue, weight loss, progressively higher and longer recurring fevers, and sometimes headache. Hepatomegaly may occur, but localizing symptoms or physical findings are often lacking. Although there is mounting evidence implicating *Bartonella* spp. in some cases of HIV-associated encephalopathy, meningoencephalitis, and neuropsychiatric disease[44–48] (see Section 4.5), lumbar puncture during acute bacteremia can be unrevealing.[6,8]

By way of contrast, *B. henselae* bacteremia in HIV-uninfected persons more often presents with abrupt onset of fever, which may persist or become relapsing. Localizing symptoms or physical findings remain unusual.[6,8,49,50] Aseptic meningitis concurrent with bacteremia has been documented at least once in an immunocompetent host.[50] *B. henselae* bacteremia can evolve into long-term asymptomatic persistence.[50]

B. elizabethae has been isolated only once as the cause of bacteremia and endocarditis.[9] Both *B. quintana* and *B. henselae* have also been found to cause endocarditis.[2,4,51–53] *B. vinsonii*, generally not considered a human pathogen, has been isolated once causing bacteremia and fever in an apparently otherwise healthy rancher from the western United States (case as yet unreported in detail).

(A subspecies of *B. vinsonii* has also been found recently to be a cause of bacteremia and endocarditis in a dog.[54] There has also been one report of isolation of a *Bartonella* strain genetically distinct from previously characterized species from blood of a kitten belonging to a man with *B. henselae* bacteremia.[55])

4.3. Bacillary Angiomatosis / Peliosis Related to *B. quintana* and *B. henselae*

Bacillary angiomatosis (BA; also referred to as epithelioid angiomatosis or bacillary epithelioid angiomatosis) is a disorder of neovascular proliferation originally described involving skin and regional lymph nodes of HIV-infected persons.[56-58] It has been demonstrated since to be able to involve a variety of internal organs including liver, spleen, bone, brain, lung, and bowel,[49,59-66] and to occur in other immunocompromised[49,61,67] as well as immunocompetent hosts.[68,69] *B. henselae* and *B. quintana* have been inculpated in BA both by direct culture[1,8,14,49,70] and by PCR amplification from tissue of specific 16 S RNA gene sequences.[14,22,67,69,71]

Cutaneous BA lesions are perhaps the most dramatic manifestation of *B. quintana* and *B. henselae* infections (Fig. 2). Such lesions often arise in crops, but both the temporal pattern of development and the gross morphologic characteristics can vary. They can be remarkably similar to lesions of verruga peruana, but as most cases have been identified in HIV-infected persons outside the region of endemic *B. bacilliformis*, the major clinical differential diagnoses are usually Kaposi's sarcoma and pyogenic granuloma. Characteristic of the gross morphology of BA skin lesions[72] are subcutaneous or dermal nodules, and/or single or multiple dome-shaped, skin-colored or red to purple papules, any of which may display ulceration, serous or bloody drainage, and crusting. Lesions can range in diameter from millimeters to centimeters, number from a few to hundreds, be fixed or freely mobile, be associated with enlargement of regional lymph nodes, involve mucosal surfaces or deeper soft tissues, occur in a variety of distributions, and bleed copiously when incised. Visceral lesions can be quite dramatic as well, both in their number and heterogeneity of gross appearance.

The histologic distinction of BA from other neovascular tumors has been clearly defined.[72,73] BA lesions consist of lobular proliferations of small blood vessels containing plump, cuboidal endothelial cells interspersed with mixed inflammatory cell infiltrates having a predominance of neutrophils. Endothelial cell atypia, mitoses, and necrosis may be present. Fibrillar- or granular-appearing amphophilic material is often present in interstitial areas when stained with hematoxylin and eosin. Warthin–Starry staining or electron microscopy demonstrates these to be clusters of bacilli.

Bacillary peliosis (BP), originally described involving the liver and sometimes spleen in HIV-infected persons,[74] has since been identified in other immunosup-

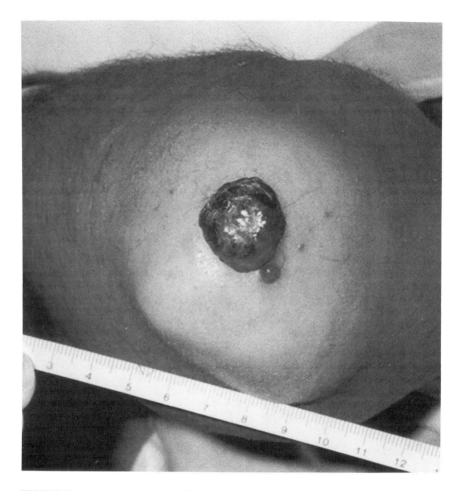

FIGURE 2. A crop of cutaneous bacillary angiomatosis lesions on the elbow of an AIDS patient. The largest lesion was of variegated purple color and had an ulcerated surface that wept serous fluid. It began a month earlier as a small cherry angiomalike lesion, much like the three adjacent smaller lesions which had all since erupted within the preceding week. All lesions involuted with doxycycline therapy.

pressed persons and found to involve lymph nodes as well.[49,75] Involved organs contain numerous blood-filled cystic structures whose size can range from microscopic to several millimeters. Pathologic findings in hematoxylin and eosin-stained tissue include partially endothelial cell-lined peliotic spaces often separated from surrounding parenchymal cells by fibromyxoid stroma containing a mixture of inflammatory cells, dilated capillaries, and clumps of granular material. Such clumps are filled with Warthin–Starry-staining bacilli.[74]

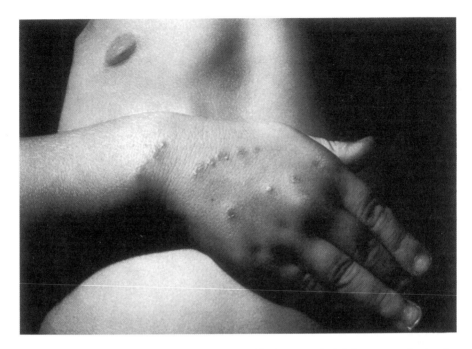

FIGURE 3. A child with typical cat-scratch disease. The image on the left demonstrates the site of the original cat scratch injuries, and the image on the right demonstrates the remarkable right axillary adenopathy that ensued. (Courtesy of V. H. San Joaquin, M. D., University of Oklahoma Health Sciences Center.)

Inflammatory reactions in immunocompromised hosts resulting from *B. henselae* infection without associated angiomatosis or peliosis have been reported involving liver, spleen, lymph nodes, heart, lung, and bone marrow. They are characterized by nodular collections of lymphocytes and nonepithelioid histiocytes which may become centrally necrotic, containing aggregates of neutrophils and karyorrhexic debris suggestive of microscopic abscess formation.[76,77] These may represent a clinical-pathologic link with CSD.[78]

4.4. Cat-Scratch Disease

Of all of the *Bartonella* species, CSD has been associated only with *B. henselae*. Evidence indicating its cardinal role includes the serologic responses of persons with CSD;[17,21,23,79] the identification of *B. henselae* in CSD lymphadenitis by culture,[24] PCR-based DNA amplification,[25–27] and immunocytochemistry;[80] detection of *B. henselae* in CSD skin test antigens by PCR;[29,30] and the recovery of *B.*

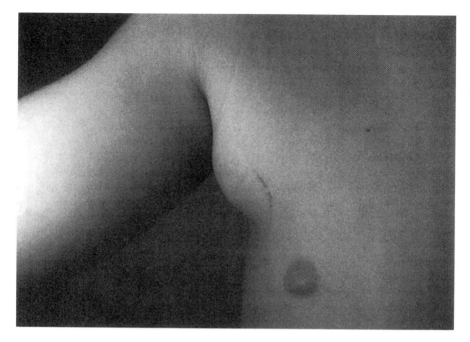

FIGURE 3. *(Continued.)*

henselae from the blood of healthy cats (which can be persistently bacteremic)[13–15] and from cat fleas.[14]

The various manifestations that comprise CSD have been recognized over the past 100 years, but the syndrome of "la maladie des griffes de chat" was not defined by Debré and collaborators until the middle of this century.[81] Furthermore, until this decade, CSD remained an infection in search of an agent. Therefore, the vast majority of cases of CSD remain defined only by clinical criteria and, in some cases, patient reactions to unstandardized skin test antigens. Although it is reasonable to ascribe the majority of CSD cases to *B. henselae* based on the numerous lines of evidence developed in recent years, one must remain open to the possibility that some cases may still be attributable to other agents such as the *Afipia* species.[82]

In typical CSD (about 89% of cases), a cutaneous papule or pustule usually develops within a week after contact with an animal (most commonly a kitten) at a site of inoculation (usually a scratch or bite) (Fig. 3).[83–85] Regional adenopathy (mostly involving head, neck, or upper extremity) develops in 1 to 7 weeks (Fig. 3). About a third of patients have fever, and about a sixth develop lymph node suppuration. The histopathology of nodes includes a mixture of nonspecific inflammatory reactions including granulomata and stellate necrosis. Bacilli may be

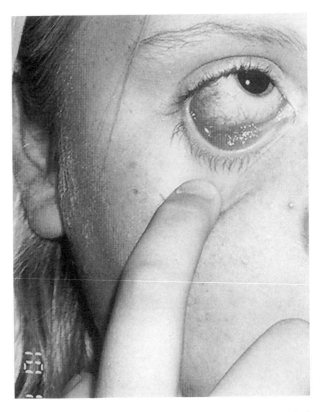

FIGURE 4. Parinaud's oculoglandular syndrome. (A) Distant view of conjunctival erythema and injection of the right eye, ipsilateral preauricular lymphadenopathy. (B) Closeup view of the same patient's lower lateral right eye, better demonstrating the granulomatous conjuctivitis.

demonstrable by Warthin–Starry or Steiner staining, and less effectively by tissue Gram staining, e.g., Brown–Hopps.

Atypical CSD (about 11%) includes Parinaud's oculoglandular syndrome (self-limited granulomatous conjunctivitis and ispilateral, usually preauricular, lymphadenitis) (Fig. 4) and various other presentations,[86] including self-limited granulomatous hepatitis/splenitis,[78,87] atypical pneumonitis,[88] osteitis, and neurologic syndromes (mainly neuroretinitis and encephalopathy).[89] Neuroretinitis manifests as fairly sudden loss of visual acuity, usually unilaterally, sometimes preceded by an influenzalike syndrome or development of unilateral lymphadenopathy. The most common retinal manifestation is papilledema associated with macular exudates in a star formation (first associated with CSD in 1984[90]), which usually has a favorable spontaneous course. Although this process can be considered characteristic of CSD neuroretinitis, it is not pathognomonic; other types of inflammation also have been reported in CSD neuroretinitis.

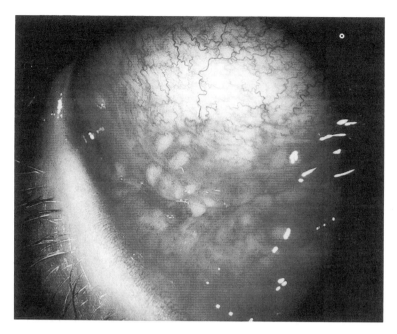

FIGURE 4. *(Continued.)*

Encephalopathy (usually manifested by headache with or without fever followed by mild to profound changes in higher cortical functions including seizures in up to half of cases, but rarely with permanent sequelae) probably occurs in 2–4% of all CSD cases recognized, although estimates range as widely as 1 to 7%.[91] Extrapolating from an estimated U.S. CSD case rate of 9.3 per year per 100,000 population,[92] 2–4% would represent between 500 and 1000 annual CSD encephalopathy cases in the United States.

The diagnosis of CSD easily can be overlooked if the clinician fails to obtain an adequate history, especially in the case of the atypical syndromes. With domestic cats representing the single largest category of companion animals in the United States, the importance of accurate history regarding animal exposure cannot be emphasized enough when evaluating a patient with findings consistent with one of these syndromes. Fortunately, in most CSD cases, whether typical or atypical, spontaneous resolution occurs in 2–4 months.

4.5. Human Immunodeficiency Virus-Associated Neurologic Syndromes

It is in the context of HIV infection that the most recent associations of *Bartonella* infections and neurologic manifestations have occurred. There is mounting evidence implicating *B. henselae* and / or *quintana* in a small proportion of cases of HIV-associated brain lesions, meningoencephalitis, encephalopathy, and neuropsychiatric disease[44-48,62] which cannot be ascribed to other causes.

Intracerebral BA was first recognized in 1990[62] in a man with AIDS who had new onset of seizures leading to the discovery of a left temporal lobe brain mass which was twice biopsied. The diagnosis was not appreciated and confirmed by the demonstration of Warthin–Starry-staining bacilli until the subsequent development and recognition of a cutaneous lesion. Erythromycin therapy caused involution of the skin and brain lesions.

In 1994, antibody and DNA amplification evidence of *Bartonella* infection was reported in the setting of neurologic manifestations complicating HIV infection.[44] Prevalence of antibodies reactive with *B. henselae* was 32% among 50 HIV-infected persons with neurologic disease, compared with less than 6% among HIV-infected persons not selected for neurologic disease. In all 16 cases identified, the CSF:serum ratio of *Bartonella*-reactive antibodies suggested antibody production in CSF, consistent with CNS infection. The enzyme immunoassay utilized had specificity for *Bartonella* spp. but not necessarily for *B. henselae*. (The *B. quintana* type strain used to demonstrate absence of cross-reactivity has since been shown to lack the antigenic qualities of contemporary *B. quintana* isolates.[48,79,93]) In the three cases in which IgM but not IgG antibodies were detected in CSF, PCR yielded amplicons of the expected size using primers specific for a 296-bp section of the *Bartonella* 16 S rRNA gene. A subsequent study evaluating autopsy brain tissue reported detection of *B. henselae* by immunofluorescence staining and by PCR amplification / sequencing of 16 S rRNA gene segment in three AIDS dementia patients with elevated CSF:serum indices of *B. henselae*-reactive antibody.[46]

A subsequent nested case control study of a geographic subpopulation of the Multicenter AIDS Cohort Study (MACS) has confirmed an association between the presence of serum anti-*Bartonella* IgM (implying recent infection) and increased risk of development of neuropsychological decline or dementia. It was estimated that at least 4% of new cases of HIV-associated dementia or neuropsychological decline could be ascribed to *Bartonella* infections, and therefore be potentially treatable with antibiotics. Subsequent case reports have added anecdotal evidence suggesting the potential utility of antimicrobial therapy in reversing *Bartonella*-associated neuropsychiatric abnormalities in HIV-infected persons.[45]

5. LABORATORY DIAGNOSIS

The laboratory diagnosis of *Bartonella*-associated diseases can be achieved through modified conventional bacteriologic culture methods,[5,24,94-96] coculture with endothelial cells,[1] immunoserologic[21,79] or immunocytochemical means,[97] and/or DNA amplification.[26,27,30,67] Approaches that are currently practical for the majority of clinical laboratories are culture and serology. Serologic testing is becoming a mainstay of diagnosis, particularly for that part of the clinical spectrum of diseases occupied by CSD and CNS infection.

5.1. Culture

Incubation of specimens of blood (after lysis ± concentration by centrifugation, such as in the Isolator™ system), tissue (after homogenization), or CSF (after concentration by cytocentrifugation) on fresh heme-containing agar media at 35–37°C in 5–10% CO_2 and elevated humidity for at least 14 days offers the most expedient recovery of isolates of non-*bacilliformis* *Bartonella* species. (*B. bacilliformis* grows best at 25–30°C.) Freshly prepared media provide optimal recovery. Heart infusion agar with 5–10% defibrinated rabbit or horse blood supports better growth of most stains than do chocolate or 5% sheep blood agars. Plates sealed after 24 hr of incubation with plastic film or shrink wraps to preserve moisture content usually can be incubated up to 30 days without notable deterioration.

Bartonella spp. generally grow best on solid or semisolid media. In broth they do not produce turbidity. Often, they do not convert enough oxidizable substrate for CO_2 detection blood culture systems to identify growth. Such approaches, including various Bactec™ systems or the equivalent, require unprompted microscopic examination of blood cultures with acridine orange staining after 7 days of incubation and subculture if bacilli are observed.[3,5] Another CO_2 detection blood culture system, Bact-Alert™, has been reported to yield positive growth algorithms in five cases of *B. henselae* bacteremia. Although Gram stains of the broth and routine 72-hr subcultures proved negative, acridine orange and Warthin–Starry staining demonstrated bacilli, and phase contrast microscopy of wet mounts revealed bacilli with "ratchety motility." Specific immunofluorescent labeling of organisms obtained directly from the broth or subsequently subcultured on semisolid media identified *B. henselae*.[96] Recent studies also have demonstrated that *B. henselae* can be isolated from clinical material using a defined liquid medium.[95] Growth of *B. henselae* in this medium occurred from three patients, including one with neuroretinitis, in 10–16 days.

B. henselae and *B. quintana* have been cultured from a variety of tissue sources, albeit infrequently, in most cases either by direct plating of tissue homogenate or aspirate[8,15,24,47] or by coculture with an endothelial cell line.[1] The coculture technique is not practical for most clinical microbiology laboratories. In the absence of coculture, more than a month of incubation has been necessary to yield evident colonies from some tissue specimens.[15,24] Because selective culture techniques have not been developed, recovery of isolates from specimens such as skin may be more difficult if indigenous or contaminating flora are present.

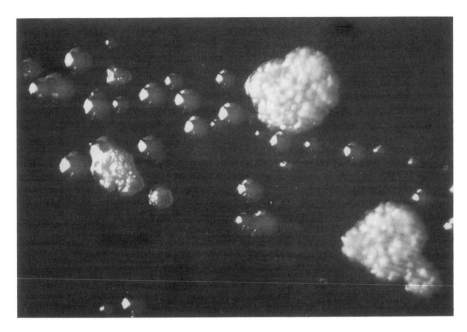

FIGURE 5. Colonies of *Bartonella henselae* demonstrating two morphologic types in close proximity on the same agar plate: irregular, raised, whitish, rough, and dry in appearance, or smaller, circular, tan, and moist in appearance.

Colonies of *Bartonella* spp. are sticky, self-adherent, and of two morphologic types: irregular, raised, whitish, rough, and dry-appearing, or smaller, circular, tan, and moist-appearing (Fig. 5). Both types may be present in the same culture. However, the degree of colonial heterogeneity varies by species. Fresh isolates of *B. henselae* typically have a greater proportion of rough colonies than *B. quintana*. Repeated subcultures cause most strains of *B. henselae* to assume predominantly smooth colony morphology.

Gram stain of a colony reveals small, gram-negative, slightly curved rods (which may mimic *Haemophilus*, *Campylobacter* or *Helicobacter*), and a wet mount usually demonstrates twitching motility. (*B. bacilliformis* possesses polar flagella whereas *B. henselae*, *B. elizabethae*, and, to lesser extent, *B. quintana* have twitching motility believed related to pili.) Presumptive identification of *B. henselae* or *B. quintana* can be made on the basis of these colony, Gram stain, and wet mount features, plus a lengthy (>7 day) period of incubation before appearance, negative catalase and oxidase reactions, and absent acid production from carbohydrates.

Additional methods may be employed to confirm the identity of isolates or isolates may be referred to a laboratory experienced with *Bartonella* spp. for confirmatory identification. A reliable means of identification is immunofluorescence using species-specific antisera,[94,96,98] but the reagents are not widely available. Cellular fatty acid composition determination by gas–liquid chromatography (GLC) is useful.[6–8,94] Commercial identification kits do not contain *Bartonella* spp. in their databases, but the Microscan™ rapid anaerobe panel distinguishes *Bartonella* spp. from species that are in its database. With careful adjustment of inoculum size, it is possible to distinguish *B. henselae* and *B. quintana* on the basis of biotype codes derived from the reactions in this panel.[94] PCR-based and DNA hybridization techniques can also be used to distinguish species.[7,8,11,26,55] Molecular subtyping of strains can also be performed using PCR-based restriction fragment length polymorphisms (RFLP)[5,99,100] and repetitive extragenic palindromic PCR (REP-PCR).[55]

5.2. Serology

As culture of *Bartonella* spp. remains technically difficult, alternative means of identification of infection are and will remain important. In mice, specific serologic responses to infection with *B. quintana* and *B. henselae* have been demonstrated by enzyme immunoassay (EIA) and immunoblot.[98] However, human antibody responses often have been substantially cross-reactive between *B. quintana* and *B. henselae* in assays so far reported.[79,93]

EIA and radioimmunoprecipitation have been found comparably more sensitive than hemagglutination and immunofluorescence assays in older studies of human antibodies to *B. quintana*.[101] With the EIA, all cases in a small series of acute primary or relapsed trench fever were found to have measurable, although often low, levels of anti-*B. quintana* antibodies.[102]

More recently, an immunofluorescence assay (IFA)[17,21,93] and several EIAs[23,79,94] have been described for *B. henselae* and *B. quintana*. To date, they have been used primarily to demonstrate anti-*Bartonella* antibodies in persons with CSD[17,21,23,79,93,103] and in some cases of HIV-associated aseptic meningitis, encephalopathy, or neuropsychiatric disease.[44,47,48] Human antibody responses often have been substantially cross-reactive between *B. quintana* and *B. henselae* in assays so far reported,[79,93] and the tests reported have undergone different degrees of scrutiny and corroboration.

The IFA and one EIA for antibodies to *B. henselae* and *B. quintana*, both of which use bacterial whole-cell antigens, have been compared with one another, with an EIA for antibodies to *A. felis*, and with findings of cat scratch antigen skin testing in evaluating patients with the clinical diagnosis of CSD.[79,93] CSD patients had no higher levels of antibodies to *A. felis* than did control persons, whereas most CSD patients had evidence of anti-*Bartonella* antibodies which were absent in controls. Detection of anti-*Bartonella* antibodies appeared to be a superior marker of CSD than did the reaction to skin test antigen. Unfortunately, neither IFA nor EIA could discriminate between anti-*B. henselae* and anti-*B. quintana* antibodies. Indeed, in many of the assays, serologic reactivity to *B. quintana* was stronger than that to *B. henselae,* despite the fact that all other lines of evidence to date point only to *B. henselae* playing an etiologic role in CSD. Likewise, EIA studies in some HIV-infected persons with encephalopathy or neuropsychiatric findings have demonstrated antibodies reactive with a formalin-fixed whole bacterial cell antigen preparation of *B. henselae* but have not satisfactorily demonstrated *B. henselae* specificity of the antibodies.[48] Alternative EIA methodology, using partially[94] or highly purified[104] antigen components, will likely prove more species-specific.

5.3. Detection in Tissue in the Absence of Culture

Bartonella organisms can be detected in tissue by silver impregnation staining (e.g., Warthin–Starry, Steiner), which is not specific, or by immunocytochemical labeling[76,80,97] or PCR amplification methods, both of which can provide species-specificity. Reagents for immunocytochemical labeling are not widely available. PCR is becoming widely used and offers potentially high sensitivity. Initial reports of this method identified sequences of amplified segments and ultimately the entirety of the *B. henselae* 16 S rRNA gene.[10,67] Subsequently, new strategies involving PCR amplification of other genomic sites and species-specific probe hybridization have simplified this approach to detection.[25,26,105] PCR amplification has also enabled identification of *B. henselae* in CSF and brain tissue of patients with HIV-related neurologic processes in the absence of culture recovery.[44,46]

6. TREATMENT OF *BARTONELLA*-RELATED CLINICAL SYNDROMES

The standard treatment for *B. bacilliformis* infection is oral chloramphenicol in a dose of 2 g/day for at least 1 week. As an alternative, oral doxycycline or other tetracyclines can be given for a comparable duration. Parenteral therapy can be substituted if oral intake or bowel absorption is impaired.

Based on the combination of ease of administration, low cost, and observed clinical effectiveness, the initial therapy of choice for uncomplicated bacteremia and processes such as BA and BP caused by non-*bacilliformis Bartonella* spp. should be oral erythromycin, doxycycline, or other tetracyclines.[106,107] As long as it is of adequate duration, such therapy appears effective for most manifestations. Exceptions may include bony or parenchymal involvement and endocarditis, for which initial parenteral therapy may be advantageous. In endocarditis, hemo-dynamic considerations may require valve replacement irrespective of the effect of antimicrobials on bacterial proliferation. For bacteremia, at least 4 weeks of therapy is indicated.[50] Longer duration treatment (8–12 weeks) is appropriate in the HIV-infected, if fever or bacteremia is persistent or recurrent in the HIV-uninfected,[47] and in the setting of endocarditis. For BA involving only the skin, experienced clinicians recommend 8–12 weeks of oral therapy.[106] Relapsing disease has been seen in both immunocompromised and immunocompetent hosts, especially but not only if therapy is terminated prematurely. For relapses occurring after adequately long initial treatment, chronic suppressive therapy with doxycycline or erythromycin should be considered.

There have been anecdotal reports of the utility of various agents in the treatment of CSD. However, the value of antibiotic therapy remains controversial.[108] Despite *in vitro* findings suggesting likely susceptibility, treatment with a variety of β-lactams (both penicillins and cephalosporins) appears largely ineffective. Among agents with possible observed utility in CSD,[108–111] responses to trimethoprim/sulfamethoxazole and fluoroquinolones in other manifestations of *Bartonella* infections have been inconsistent. The use of rifampin and gentamicin in these areas requires further investigation.

In the specific context of neurologic manifestations of *B. henselae* (and possibly *quintana*) infections, in light of the increasing evidence of associated bacteremia and/or primary infection within the CNS cited above, antimicrobial therapy is probably prudent. Parenteral therapy is not commonly indicated, and oral erythromycin or tetracyclines should probably be agents of first choice, administered for several months. The high CNS penetration of rifampin and ofloxacin would make them reasonable candidates as adjunctive (rifampin) or alternative (ofloxacin) agents. Further studies of these issues may be facilitated by improved techniques for the diagnosis of *Bartonella* infections.

7. FUTURE DEVELOPMENTS

Our understanding of the extent and ramifications of human infection by members of the genus *Bartonella* is undergoing a rebirth. Long-recognized clinical syndromes (CSD, bacteremia, "culture-negative" endocarditis) have been associated with newly recognized human pathogens (*B. henselae, B. elizabethae*). Newly recognized clinical syndromes (BA and BP) have been associated with long-identified (*B. quintana*) and recently discovered (*B. henselae*) agents. Improved culture and nonculture diagnostic techniques, coupled with the suspicion of the astute clinician, will provide the tools to further define the full spectrum of *Bartonella* infections. Additional species may be found to infect humans, and new syndromes may be described. The full enzootic distribution of *Bartonella* species recognized to be (perhaps incidental) human pathogens will require exploration.

The mechanisms by which *Bartonella* species cause their clinical syndromes, some of which are unique, will also demand further elucidation.

REFERENCES

1. Koehler, J. E., Quinn, F. D., Berger, T. G., LeBoit, P. E., and Tappero, J. W., 1992, Isolation of *Rochalimaea* species from cutaneous and osseous lesions of bacillary angiomatosis, *N. Engl. J. Med.* **327:**1625–1632.
2. Spach, D. H., Callis, K. P., Paauw, D. S., Houze, Y. B., Schoenknecht, F. D., Welch, D. F., Rosen, H., and Brenner, D. J., 1993, Endocarditis caused by *Rochalimaea quintana* in a patient infected with human immunodeficiency virus, *J. Clin. Microbiol.* **31:**692–694.
3. Larson, A. M., Dougherty, M. J., Nowowiejski, D. J., Welch, D. F., Matar, G. M., Swaminathan, B., and Coyle, M. B., 1994, Detection of *Bartonella (Rochalimaea) quintana* by routine acridine orange staining of broth blood cultures, *J. Clin. Microbiol.* **32:**1492–1496.
4. Drancourt, F., Etienne, J., Goldstein, F., Acar, J., and Raoult, D., 1995, *Bartonella (Rochalimaea) quintana* endocarditis in three homeless men, *N. Engl. J. Med.* **332:**419–423.
5. Spach, D. H., Kanter, A. S., Dougherty, M. J., Larson, A. M., Coyle, M. B., Brenner, D. J., Swaminathan, B., Matar, G. M., Welch, D. F., Root, R. K., and Stamm, W. E., 1995, *Bartonella (Rochalimaea) quintana* bacteremia in inner-city patients with chronic alcoholism, *N. Engl. J. Med.* **332:**424–428.
6. Slater, L. N., Welch, D. F., Hensel, D., and Coody, D. W., 1990, A newly recognized fastidious gram-negative pathogen as a cause of fever and bacteremia, *N. Engl. J. Med.* **323:**1587–1593.
7. Regnery, R. L., Anderson, B. E., Clarridge, J. E., III, Rodriguez-Barradas, M. C., Jones, D. C., and Carr, J. H., 1992, Characterization of a novel *Rochalimaea* species, *R. henselae* sp. nov., isolated from blood of a febrile, human immunodeficiency virus-positive patient, *J. Clin. Microbiol.* **30:**265–274.
8. Welch, D. F., Pickett, D. A., Slater, L. N., Steigerwalt, A. G., and Brenner, D. J., 1992, *Rochalimaea henselae* sp. nov., a cause of septicemia, bacillary angiomatosis, and parenchymal bacillary peliosis, *J. Clin. Microbiol.* **30:**275–280.
9. Daly, J. S., Worthington, M. G., Brenner, D. J., Moss, C. W., Hollis, D. G., Weyant, R. S., Steigerwalt, A. G., Weaver, R. E., Daneshvar, M. I., and O'Connor, S. P., 1993, *Rochalimaea elizabethae* sp. nov. isolated from a patient with endocarditis, *J. Clin. Microbiol.* **31:**872–881.
10. Relman, D. A., Lepp, P. W., Sadler, K. N., and Schmidt, T. M., 1992, Phylogenetic relationships among the agent of bacillary angiomatosis, *Bartonella bacilliformis,* and other alpha-proteobacteria, *Mol. Microbiol.* **6:**1801–1807.
11. Brenner, D. J., O'Connor, S. P., Winkler, H. H., and Steigerwalt, A. G., 1993, Proposals to unify the genera *Bartonella* and *Rochalimaea,* with descriptions of *Bartonella quintana* comb. nov., *Bartonella vinsonii* comb. nov., *Bartonella henselae* comb. nov., and *Bartonella elizabethae* comb. nov., and to remove the family *Bartonellaceae* from the order *Rickettsiales, Int. J. Syst. Bacteriol.* **43:**777–786.
12. Birtles, R. J., Harrison, T. G., Saunders, N. A., and Molyneux, D. H., 1995, Proposals to unify the genera *Grahamella* and *Bartonella,* with descriptions of *Bartonella talpae* comb. nov., *Bartonella peromysci* sp. nov., *Bartonella taylorii* sp. nov., and *Bartonella doshiae* sp. nov., *Int. J. Syst. Bacteriol.* **45:**1–8.
13. Regnery, R., Martin, M., and Olson, J., 1992, Naturally occurring "*Rochalimaea henselae*" infection in domestic cat [letter], *Lancet* **340:**557–558.
14. Koehler, J. E., Glaser, C. A., and Tappero, J. W., 1994, *Rochalimaea henselae* infection: A new zoonosis with the domestic cat as reservoir, *J. Am. Med. Assoc.* **271:**531–535.

15. Demers, D. M., Bass, J. W., Vincent, J. M., Person, D. A., Noyes, D. K., Staege, C. M., Samslaska, C. P., Lockwood, N. H., Regnery, R. L., and Anderson, B. E., 1995, Cat scratch disease in Hawaii: Etiology and seroepidemiology, *J. Pediatr.* **127:**23-26.
16. Chomel, B. B., Abbot, R. C., Kasten, R. W., Floyd-Hawkins, K. A., Kass, P. H., Glaser, C. A., Pedersen, N., and Koehler, J. E., 1995, *Bartonella henselae* prevalence in domestic cats in California: Risk factors and association between bacteremia and antibody titers, *J. Clin. Microbiol.* **33:**2445-2450.
17. Zangwill, K. M., Hamilton, D. H., Perkins, B. A., Regnery, R. L., Plikaytis, B. D., Hadler, J. L., Cartter, M. L., and Wenger, J. D., 1993, Cat scratch disease in Connecticut. Epidemiology, risk factors, and evaluation of a new diagnostic test, *N. Engl. J. Med.* **329:**8-13.
18. Childs, J. E., Rooney, J. A., Cooper, J. L., Olson, J. G., and Regnery, R. L., 1994, Epidemiologic observations on infection with *Rochalimaea* species among cats living in Baltimore, Md, *J. Am. Vet. Med. Assoc.* **204:**1775-1778.
19. Jameson, P., Greene, C., Regnery, R., Dryden, M., Marks, A., Brown, J., Cooper, J., Glaus, B., and Greene, R., 1995, Prevalence of *Bartonella henselae* antibodies in pet cats throughout regions of North America, *J. Infect. Dis.* **172:**1145-1149.
20. Ueno, H., Muramatsu, Y., Chomel, B. B., Hohdatsu, T., Koyama, H., and Morita, C., 1995, Seroepidemiological survey of *Bartonella (Rochalimaea) henselae* in domestic cats in Japan, *Microbiol. Immunol.* **39:**339-341.
21. Regnery, R. L., Olson, J. G., Perkins, B. A., and Bibb, W., 1992, Serologic response to "*Rochalimaea henselae*" antigen in suspected cat-scratch disease, *Lancet* **339:**1443-1445.
22. Tappero, J. W., Mohle-Boetani, J., Koehler, J., Swaminathan, B., Berger, T. G., LeBoit, P. E., Smith, L. L., Wenger, J. D., Pinner, R. W., Kemper, C., and Reingold, A. L., 1993, The epidemiology of bacillary angiomatosis and bacillary peliosis, *J. Am. Med. Assoc.* **269:**770-775.
23. Barka, N. R., Hadfield, T., Patnaik, M., Schwartzman, W. A., and Peter, J. B., 1993, EIA for detection of *Rochalimaea henselae*-reactive IgG, IgM, and IgA antibodies in patients with suspected cat scratch disease [letter], *J. Infect. Dis.* **167:**1503-1504.
24. Dolan, M. J., Wong, M. T., Regnery, R. L., Jorgensen, J. H., Garcia, M., Peters, J., and Drehner, D., 1993, Syndrome of *Rochalimaea henselae* adenitis suggesting cat scratch disease, *Ann. Intern. Med.* **118:**331-336.
25. Waldvogel, K., Regnery, R. L., Anderson, B. E., Caduff, R., Caduff, J., and Nadal, D., 1994, Disseminated cat-scratch disease: Detection of *Rochalimaea henselae* in affected tissue, *Eur. J. Pediatr.* **153:**23-27.
26. Anderson, B., Sims, K., Regnery, R., Robinson, L., Schmidt, M. J., Goral, S., Hager, C., and Edwards, K., 1994, Detection of *Rochalimaea henselae* DNA in specimens from cat-scratch disease patients by PCR, *J. Clin. Microbiol.* **32:**942-948.
27. Goral, S., Anderson, B., Hager, C., and Edwards, K., 1994, Detection of *Rochalimaea henselae* DNA by polymerase chain reaction from suppurative nodes of children with cat-scratch disease, *Pediatr. Infect. Dis. J.* **13:**994-997.
28. Le, H. H., Palay, D. A., Anderson, B., and Steinberg, J. P., 1994, Conjunctival swab to diagnose ocular cat scratch disease, *Am. J. Ophthalmol.* **118:**249-250.
29. Perkins, B. A., Swaminathan, B., Jackson, L. A., Brenner, D. J., Wenger, J. D., Regnery, R. L., and Wear, D. J., 1992, Case 22-1992. Pathogenesis of cat scratch disease [letter], *N. Engl. J. Med.* **327:**1599-1600.
30. Anderson, B., Kelly, C., Threlkel, R., and Edwards, K., 1993, Detection of *Rochalimaea henselae* in cat-scratch disease skin test antigens, *J. Infect. Dis.* **168:**1034-1036.
31. Ricketts, W. E., 1947, Carrion's disease. A study of the incubation period in thirteen cases, *Am. J. Trop. Med.* **27:**657-659.
32. Strong, R. P., Tyzzer, E. E., Brues, C. T., Sellards, A. W., and Gastiaburu, J. C., 1915, *Report of the First Expedition to South America, 1913,* Harvard University Press, Cambridge, MA.

33. Ricketts, W. E., 1948, *Bartonella bacilliformis* anemia (Oroya fever), *Blood* **3:**1025–1049.
34. Reynafarje, C., and Ramos, J., 1961, The hemolytic anemia of human bartonellosis, *Blood* **17:**562–578.
35. Ricketts, W. E., 1949, Clinical manifestations of Carrión's disease, *Arch. Intern. Med.* **84:**751–781.
36. Cuadra, M., 1956, Salmonellosis complication in human bartonellosis, *Tex. Rep. Biol. Med.* **14:**97–113.
37. Pinkerton, H., and Weinman, D., 1940, Toxoplasma infection in man, *Arch. Pathol.* **30:**374–392.
38. Garcia-Caceres, U., and Garcia, F. U., 1991, Bartonellosis. An immunosuppressive disease and the life of Daniel Alcides Carrion, *Am. J. Clin. Pathol.* **95**(Suppl. 1)**:**S56–S66.
39. Dooley, J. R., 1976, Baronellosis, in: *Pathology of Tropical and Extraordinary Diseases* (C. H. Binford and D. H. Connored, eds.), Armed Forces Institute of Pathology, Washington, DC, pp. 190–193.
40. Arias-Stella, J., Lieberman, P. H., Erlandson, R. A., and Arias-Stella, J., Jr., 1986, Histology, immunohistochemistry, and ultrastructure of the verruga in Carrion's disease, *Am. J. Surg. Pathol.* **10:**595–610.
41. Jackson, L. A., and Spach, D. H., 1996, Emergence of *Bartonella quintana* infection among homeless persons, *Emerg. Infect. Dis.* **2:**141–144.
42. McNee, J. W., and Renshaw, A., 1916, "Trench fever": A relapsing fever occurring with the British forces in France, *Br. Med. J.* **1:**225–234.
43. Liu, W.-T., 1984, Trench fever: A resumé of literature and a note on some obscure phases of the disease, *Chin. Med. J.* **97:**179–190.
44. Schwartzman, W. A., Patnaik, M., Barka, N. E., and Peter, J. B., 1994, *Rochalimaea* antibodies in HIV-associated neurologic disease, *Neurology* **44:**1312–1316.
45. Baker, J., Ruiz-Rodriguez, R., Whitfield, M., Heon, V., and Berger, T. G., 1995, Bacillary angiomatosis: A treatable cause of acute psychiatric symptoms in human immunodeficiency virus infection. *J. Clin. Psychiatry* **56:**161–166.
46. Patnaik, M., Schwartzman, W. A., and Peter, J. B., 1995, *Bartonella henselae:* Detection in brain tissue of patients with AIDS-associated neurological disease, *J. Invest. Med.* **43:**368A (abstract of poster presentation at the 1995 Clinical Research Meeting, San Diego).
47. Wong, M. T., Dolan, M. J., Lattuada, C. P., Jr., Regnery, R. L., Garcia, M. L., Mokulis, E. C., LaBarre, R. C., Ascher, D. P., Delmar, J. A., Kelly, J. W., Leigh, D. R., McRae, A. C., Reed, J. B., Smith, R. E., and Melcher, G. P., 1995, Neuroretinitis, aseptic meningitis, and lymphadenitis associated with *Bartonella (Rochalimaea) henselae* infection in immunocompetent patients and patients infected with human immunodeficiency virus type 1, *Clin. Infect. Dis.* **21:**352–360.
48. Schwartzman, W. A., Patnaik, M., Angulo, F. J., Visscher, B. R., and Peter, J. B., 1995, *Bartonella (Rochalimaea)* antibodies, dementia, and cat ownership in human immunodeficiency virus-infected men, *Clin. Infect. Dis.* **21:**954–959.
49. Slater, L. N., Welch, D. F., and Min, K.-W., 1992, *Rochalimaea henselae* causes bacillary angiomatosis and peliosis hepatis, *Arch. Intern. Med.* **152:**602–606.
50. Lucey, D., Dolan, M. J., Moss, C. W., Garcia, M., Hollis, D. G., Wegner, S., Morgan, G., Almeida, R., Leong, D., Greisen, K. S., Welch, D. F., and Slater, L. N., 1992, Relapsing illness due to *Rochalimaea henselae* in normal hosts: Implication for therapy and new epidemiologic associations, *Clin. Infect. Dis.* **14:**683–688.
51. Jalava, J., Kotilainen, P., Nikkari, S., Skurnik, M., Vänttinen, E., Lehtonen, O.-P., Eerola, E., and Toivanen, P., 1995, Use of polymerase chain reaction and DNA sequencing for detection of *Bartonella quintana* in the aortic valve of a patient with culture-negative infective endocarditis, *Clin. Infect. Dis.* **21:**891–896.

52. Hadfield, T. L., Warren, R., Kass, M., and Levy, C., 1993, Endocarditis caused by *Rochalimaea henselae*, *Hum. Pathol.* **24:**1140–1141.

53. Holmes, A. H., Greeough, T. C., Balady, G. J., Regnery, R. L., Anderson, B. E., O'Keane, J. C., Fonger, J. D., and McCrone, E. L., 1995, *Bartonella henselae* endocarditis in an immunocompetent adult, *Clin. Infect. Dis.* **21:**1004–1007.

54. Breitschwerdt, E. B., Kordick, D. L., Malarkey, D. E., Keene, B., Hadfield, T. L., and Wilson, K., 1995, Endocaritis in a dog due to infection with a novel *Bartonella* subspecies, *J. Clin. Microbiol.* **33:**154–160.

55. Clarridge, J. E., III, Raich, T. J., Pirwani, D., Simon, B., Tsai, L., Rodriguez-Barradas, M. C., Regnery, R., Zollo, A., Jones, D. C., and Rambo, C., 1995, Strategy to detect and identify *Bartonella* species in a routine clinical laboratory yields *Bartonella henselae* from human immunodeficiency virus-infected patient and unique *Bartonella* strain from his cat, *J. Clin. Microbiol.* **33:**2107–2113.

56. Stoler, M. H., Bonfiglio, T. A., Steigbigel, R. T., and Pereira, M., 1983, An atypical subcutaneous infection associated with acquired immune deficiency syndrome, *Am. J. Clin. Pathol.* **80:**714–718.

57. Cockerell, C. J., Webster, G. F., Whitlow, M. A., and Friedman-Kien, A. E., 1987, Epithelioid angiomatosis: A distinct vascular disorder in patients with the acquired immunodeficiency syndrome or AIDS-related complex, *Lancet* **2:**6544–6546.

58. LeBoit, P. E., Egbert, B. M., Stoler, M. H., Strauchen, J. A., Wear, D. J., Berger, T. G., Yen Benedict, T. S., Bonfiglio, T. A., and English, C. K., 1988, Epithelioid haemangioma-like vascular proliferation in AIDS: Manifestation of cat scratch disease bacillus infection? *Lancet* **1:**960–963.

59. Koehler, J. E., LeBoit, P. E., Egbert, B. M., and Berger, T. G., 1988, Cutaneous vascular lesions and disseminated cat-scratch disease in patients with the acquired immunodeficiency syndrome (AIDS) and AIDS-related complex, *Ann. Intern. Med.* **109:**449–455.

60. Milam, M. W., Balerdi, M. J., Toney, J. F., Foulis, P. R., Milam, C. P., and Behnke, R. H., 1990, Epithelioid angiomatosis secondary to disseminated cat scratch disease involving the bone marrow and skin in a patient with acquired immune deficiency syndrome: A case report, *Am. J. Med.* **88:**180–183.

61. Kemper, C. A., Lombard, C. M., Deresinski, S. C., and Tompkins, L. S., 1990, Visceral bacillary epithelioid angiomatosis: Possible manifestations of disseminated cat scratch disease in the immunocompromised host: a report of two cases, *Am. J. Med.* **89:**216–222.

62. Spach, D. H., Panther, L. A., Thorning, D. R., Dunn, J. E., Plorde, J. J., and Miller, R. A., 1992, Intracerebral bacillary angiomatosis in a patient infected with the human immunodeficiency virus, *Ann. Intern. Med.* **116:**740–742.

63. Koehler, J. E., and Cederberg, L., 1995, Intraabdominal mass associated with gastrointestinal hemorrhage: A new manifestation of bacillary angiomatoisis, *Gastroenterology* **109:**2011–2014.

64. Coche, E., Beigelman, C., Lucidarme, O., Finet, J. F., Bakdach, H., and Grenier, P., 1995, Thoracic bacillary angiomatosis in a patient with AIDS, *Am. J. Roentgenol.* **165:**56–58.

65. Mohle-Boetani, J. C., Koehler, J. E., Berger, T. G., LeBoit, P. E., Kemper, C. A., Reingold, A. L., Plikaytis, B. D., Wenger, J. D., and Tappero, J. W., 1996, Bacillary angiomatosis and bacillary peliosis in patients infected with the human immunodeficiency virus: Clinical characteristics in a case control study, *Clin. Infect. Dis.* **22:**794–800.

66. Huh, Y. B., Rose, S., Schoen, R. E., Hunt, S., Whitcomb, D. C., and Finkelstein, S., 1996, Colonic bacillary angiomatosis, *Ann. Intern. Med.* **124:**735–737.

67. Relman, D. A., Loutit, J. S., Schmidt, T. M., Falkow, S., and Tompkins, L. S., 1990, The agent of bacillary angiomatosis: An approach to the identification of uncultured pathogens, *N. Engl. J. Med.* **323:**1573–1580.

68. Cockerell, C. J., Bergstresser, P. R., Myrie-Williams, C., and Tierno, P. M., 1990, Bacillary

epithelioid angiomatosis occurring in an immunocompetent individual, *Arch. Dermatol.* **126:**787–790.

69. Tappero, J. W., Koehler, J. E., Berger, T. G., Cockerell, C. J., Lee, T.-H., Busch, M. P., Stites, D. P., Moehle-Boetani, J., Reingold, A. L., and LeBoit, P. E., 1993, Bacillary angiomatosis and bacillary splenitis in immunocompetent adults, *Ann. Intern. Med.* **118:**363–365.

70. Cockerell, C. J., Tierno, P. M., Friedman-Kien, A. E., and Kim, K. S., 1991, Clinical, histologic, microbiologic, and biochemical characterization of the causative agent of bacillary (epithelioid) angiomatosis: A rickettsial illness with features of bartonellosis, *J. Invest. Dermatol.* **97:**812–817.

71. Relman, D. A., Falkow, S., LeBoit, P. E., Perkocha, L. A., Min, K.-W., Welch, D. F., and Slater, L. N., 1991, The organism causing bacillary angiomatosis, peliosis hepatis, and fever and bacteremia in immunocompromised patients [letter], *N. Engl. J. Med.* **324:**1514.

72. Cockerell, C. J., and LeBoit, P. E., 1990, Bacillary angiomatosis: A newly characterized, pseudoneoplastic, infectious, cutaneous vascular disorder, *J. Am. Acad. Dermatol.* **22:**501–512.

73. LeBoit, P. E., Berger, T. G., Egbert, B. M., Beckstead, J. H., Yen, Y. S. B., and Stoler, M. H., 1989, Bacillary angiomatosis. The histology and differential diagnosis of a pseudoneoplastic infection in patients with human immunodeficiency virus disease, *Am. J. Surg. Pathol.* **13:**909–920.

74. Perkocha, L. A., Geaghan, S. M., Yen, T. S. B., Nishimura, S. L., Chan, S. P., Garcia-Kennedy, R., Honda, G., Stoloff, A. C., Klein, H. Z., Goldman, R. L., Meter, S. V., Farrell, L. D., and LeBoit, P. E., 1990, Clinical and pathological features of bacillary peliosis hepatitis in association with human immunodeficiency virus infection, *N. Engl. J. Med.* **323:**1581–1586.

75. Leong, S. S., Cazen, R. A., Yu, G. S. M., LeFevre, L., and Carson, J. W., 1992, Abdominal visceral peliosis associated with bacillary angiomatosis: Ultrastructural evidence of endothelial cell destruction by bacilli, *Arch. Pathol. Lab. Med.* **116:**866–871.

76. Slater, L. N., Pitha, J. V., Herrera, L., Hughson, M. D., Min, K.-W., and Reed, J. A., 1994, *Rochalimaea henselae* infection in AIDS causing inflammatory disease without angiomatosis or peliosis: Demonstration by immunocytochemistry and corroboration by DNA amplification, *Arch. Pathol. Lab. Med.* **118:**33–38.

77. Caniza, M. A., Granger, D. L., Wilson, K. H., Washington, M. K., Kordick, D. L., Frush, D. P., and Blitchington, R. B., 1995, *Bartonella henselae:* Etiology of pulmonary nodules in a patient with depressed cell-mediated immunity, *Clin. Infect. Dis.* **20:**1505–1511.

78. Liston, T. E., and Koehler, J. E., 1996, Granulomatous hepatitis and necrotizing splenitis due to *Bartonella henselae* in a patient with cancer: Case report and review of hepatosplenic manifestations of Bartonella infection, *Clin. Infect. Dis.* **22:**951–957.

79. Szelc-Kelly, C. M., Goral, S., Perez-Perez, G. I., Perkins, B. A., Regnery, R. L., and Edwards, K. M., 1995, Serologic responses to *Bartonella* and *Afipia* antigens in patients with cat scratch disease, *Pediatrics* **96:**1137–1142.

80. Min, K.-W., Reed, J. A., Welch, D. F., and Slater, L. N., 1994, Morphologically variable bacilli of cat scratch disease are identified by immunocytochemical labeling with antibodies to *Rochalimaea henselae, Am. J. Clin. Pathol.* **101:**607–610.

81. Debré, R., Lamy, M., Jammet, M. L., Costil, L., and Mozziconacci, P., 1950, La maladie des griffes de chat, *Sem. Hop. Paris* **26:**1895–1904.

82. Brenner, D. J., Hollis, D. G., Moss, C. W., English, C. K., Hall, G. S., Judy, V., Radosevic, J., Birkness, K. A., Bibb, W. F., Quinn, F. D., Swaminathan, B., Weaver, R. E., Reeves, M. W., O'Connor, S. P., Hayes, P. S., Tenover, F. C., Steigerwalt, A. G., Perkins, B. A., Daneshvar, M. I., Hill, B. C., Washington, J. A., Woods, T. C., Hunter, S. B., Hadfield, T. L., Ajello, G. W., Kaufman, A. F., Wear, D. J., and Wenger, J. D., 1991, Proposal of *Afipia* gen. nov., with *Afipia felis* sp. nov. (formerly the cat scratch disease bacillus), *Afipia clevelandensis* sp. nov. (formerly the

Cleveland Clinic Foundation strain), *Afipia broomeae* sp. nov., and three unnamed genospecies, *J. Clin. Microbiol.* **29:**2450–2460.

83. Carithers, H. A., 1985, Cat-scratch disease. An overview based on a study of 1,200 patients, *Am. J. Dis. Child.* **139:**1124–1133.

84. Moriarty, R., and Margileth, A., 1987, Cat scratch disease, *Infect. Dis. Clin. North Am.* **1:**575–590.

85. Margileth, A. M., 1993, Cat scratch disease, *Adv. Pediatr. Infect. Dis.* **8:**1–21.

86. Margileth, A. M., Wear, D. J., and English, C. K., 1987, Systemic cat scratch disease: Report of 23 patients with prolonged or recurrent severe bacterial infection, *J. Infect. Dis.* **155:**390–402.

87. Delahoussaye, P. M., and Osborne, B. M., 1990, Cat-scratch disease presenting as abdominal visceral granulomas, *J. Infect. Dis.* **161:**71–78.

88. Abbasi, S., and Chesney, P. J., 1995, Pulmonary manifestations of cat scratch disease; a case report and review of the literature, *Pediatr. Infect. Dis. J.* **14:**547–548.

89. Carithers, H., and Margileth, A., 1991, Cat scratch disease. Acute encephalopathy and other neurologic manifestations, *Am. J. Dis. Child.* **145:**98–101.

90. Dreyer, R. F., Hopen, G., Gass, D. M., and Smith, J. L., 1984, Leber's idiopathic stellate neuroretinitis, *Arch. Ophthalmol.* **102:**1140–1145.

91. Centers for Disease Control and Prevention, 1994, Encephalitis associated with cat scratch disease—Broward and Palm Beach Counties, Florida, 1994, *MMWR* **43:**915–916.

92. Jackson, L. A., Perkins, B. A., and Wenger, J. D., 1993, Cat scratch disease in the United States: An analysis of three national databases, *Am. J. Public Health* **83:**1707–1711.

93. Dalton, M. J., Robinson, L. E., Cooper, J., Regnery, R. L., Olson, J. G., and Childs, J. E., 1995, Use of *Bartonella* antigens for the serologic diagnosis of cat-scratch disease at a national referral center, *Arch. Intern. Med.* **155:**1670–1676.

94. Welch, D. F., Hensel, D. M., Pickett, D. A., San Joaquin, V. H., Robinson, A., and Slater, L. N., 1993, Bacteremia due to *Rochalimaea henselae* in a child: Practical identification of isolates in the clinical laboratory, *J. Clin. Microbiol.* **31:**2381–2386.

95. Wong, M. T., Thornton, D. C., Kennedy, R. C., and Dolan, M. J., 1995, A chemically defined medium that supports primary isolation of *Rochalimaea (Bartonella) henselae* from blood and tissue specimens, *J. Clin. Microbiol.* **33:**742–744.

96. Tierno, P. M., Jr., Inglima, K., and Parisi, M. T., 1995, Detection of *Bartonella (Rochalimaea) henselae* bacteremia using BacT/Alert blood culture system, *Am. J. Clin. Pathol.* **104:**530–536.

97. Reed, J., Brigati, D. J., Flynn, S. D., McNutt, N. S., Min, K.-W., Welch, D. F., and Slater, L. N., 1992, Immunocytochemical identification of *Rochalimaea henselae* in bacillary (epithelioid) angiomatosis, parenchymal bacillary peliosis, and persistent fever with bacteremia, *Am. J. Surg. Pathol.* **16:**650–657.

98. Slater, L. N., Coody, D. W., Woolridge, L. K., and Welch, D. F., 1992, Murine antibody responses distinguish *Rochalimaea henselae* from *Rochalimaea quintana*, *J. Clin. Microbiol.* **30:**1722–1727.

99. Matar, G. M., Swaminathan, B., Hunter, S. B., Slater, L. N., and Welch, D. F., 1993, Polymerase chain reaction-based restriction fragment length polymorphism analysis of a fragment of the ribosomal operon from *Rochalimaea* species for subtyping, *J. Clin. Microbiol.* **31:**1730–1734.

100. Roux, V., and Raoult, D., 1995, *Inter- and intraspecies identification of Bartonella (Rochalimaea) species*, *J. Clin. Microbiol.* **33:**1573–1576.

101. Herrmann, J. E., Hollingdale, M. R., Collins, M. F., and Vinson, J. W., 1977, Enzyme immunoassay and radioimmunoprecipitation tests for the detection of antibodies to *Rochalimaea (Rickettsia) quintana* (39655), *Proc. Soc. Exp. Biol. Med.* **154:**285–288.

102. Hollingdale, M. R., Herrmann, J. E., and Vinson, J. W., 1978, Enzyme immunoassay of antibody to *Rochalimaea quintana*: Diagnosis of trench fever and serologic cross-reactions among other rickettsiae, *J. Infect. Dis.* **137:**578–582.

103. Golnik, K. C., Marotto, M. E., Fanous, M. M., Heitter, D., King, L. P., Halpern, J. I., and Holley, H. P., Jr., 1994, Ophthalmic manifestations of *Rochalimaea* species, *Am. J. Ophthalmol.* **118:**145–151.

104. Anderson, B., Lu, E., Jones, D., and Regnery, R., 1995, Characterization of 17-kilodalton antigen of *Bartonella henselae* reactive with sera from patients with cat scratch disease, *J. Clin. Microbiol.* **33:**2358–2365.

105. Bergmans, A. M. C., Groothedde, J.-W., Schellekens, J. F. P., van Embden, J. D. A., Ossewaarde, J. M., and Schouls, L. M., 1995, Etiology of cat scratch disease: Comparison of polymerase chain reaction detection of *Bartonella* (formerly *Rochalimaea*) and *Afipia felis* DNA with serology and skin tests, *J. Infect. Dis.* **171:**916–923.

106. Koehler, J. E., and Tappero, J. W., 1993, Bacillary angiomatosis and bacillary peliosis in patients infected with human immunodeficiency virus, *Clin. Infect. Dis.* **17:**612–624.

107. Regnery, R. L., Childs, J. E., and Koehler, J. E., 1995, *Infections associated with Bartonella species in persons infected with the human immunodeficiency virus, Clin. Infect. Dis.* **21:**S94–S98.

108. Margileth, A. M., 1992, Antibiotic therapy for cat-scratch disease: Clinical study of therapeutic outcome in 268 patients and a review of the literature, *Pediatr. Infect. Dis. J.* **11:**474–478.

109. Bogue, C., Wise, J. D., Gray, G. F., and Edwards, K. M., 1989, Antibiotic therapy for cat-scratch disease? *J. Am. Med. Assoc.* **262:**813–816.

110. Holley, H. P., Jr., 1991, Successful treatment of cat-scratch disease with ciprofloxacin, *J. Am. Med. Assoc.* **265:**1563–1565.

111. Collipp, P. J., 1992, Cat-scratch disease: Therapy with trimethoprim–sulfamethoxazole, *Am. J. Dis. Child.* **146:**397–399.

12

Virulence Determinants of
Bartonella bacilliformis

MICHAEL F. MINNICK

1. INTRODUCTION

Bartonella bacilliformis is a facultative intracellular bacterial parasite of human erythrocytes and endothelial cells. Carrión's disease, Oroya fever and verruga peruana are all terms describing the pathological consequences of a human infection with *B. bacilliformis*. Although infections involving *Bartonella* species such as *B. henselae* and *B. quintana* occur worldwide, Carrión's disease is uniquely endemic to South America. *B. bacilliformis* infections are a health problem in many rural areas of South America[1] and to travelers who visit these regions.[2] Outbreaks of bartonellosis have been reported in the mountainous regions of Peru, Colombia, Ecuador, Bolivia, Chile, and Guatemala.[3] A recent epidemic in 1987 was blamed for 14 deaths and 14 additional cases in the Peruvian village of Shumpillan.[4] In some areas of Peru, up to 60% of the local population are seropositive for antibartonella antibodies,[5] and chronic asymptomatic carrier rates are as high as 15%.[6] The possibility of alternative reservoirs of *B. bacilliformis* such as plants or other animals has never been studied.

The phlebotamine sandfly, *Lutzomyia verrucarum*, is the vector for transmission of the bacterium to humans. The female sandfly transmits the pathogen during nocturnal blood feeding on humans. Presumably, the insect feeds on the blood of an infected individual and spreads the pathogen via her saliva during a subse-

MICHAEL F. MINNICK • Division of Biological Sciences, The University of Montana, Missoula, Montana 59812-1002.

Rickettsial Infection and Immunity, edited by Anderson *et al.* Plenum Press, New York, 1997.

quent blood meal. It is not known if *B. bacilliformis* replicates in or infects the sandfly vector, but either of these capabilities would certainly enhance transmission of the bacterium to humans. It is also not known if alternative vectors of *B. bacilliformis* exist, although a recent review suggests that sandflies that are closely related to *L. verrucarum* may also transmit the pathogen to humans.[3] Like *B. quintana*,[7] transmission of *B. bacilliformis* by bloodsucking arthropods such as lice or fleas might be possible and would be enhanced by conditions of poor hygiene and overcrowding. The bacterium is not contagious between humans.[8]

Treatment of the disease has been fostered by the advent of antibiotics. Prior to antimicrobials, bartonella infections were fatal nearly 40% of the time.[9,10] Fortunately, *B. bacilliformis* is sensitive to most antimicrobials including penicillin, tetracycline, chloramphenicol, and various aminoglycosides.[8]

Numerous virulence determinants have been implicated in the virulence of *B. bacilliformis* including motility, various adhesins, deformin, β-hemolysin, and the invasion locus *ialAB*. This chapter reviews the present data regarding these virulence factors.

2. ESTABLISHMENT OF INFECTION

2.1. Biphasic Disease Manifestations

Bartonellosis is an unusual disease in that it presents with two distinct clinical and temporal phases: a primary (hematic) phase and a secondary (tissue) phase. The biphasic nature of the disease confused many early investigators who believed that the two stages were caused by different etiologic agents. Daniel Carrión definitively demonstrated that both phases are actually produced by the same pathogen when he injected himself with scraping of verruga peruana and subsequently died from a case of Oroya fever.[1] In 1926, Noguchi satisfied Koch's postulates using infected macaque monkeys.[11]

Humans present with symptoms of the primary (hematic) stage of bartonellosis within 2 to 3 weeks following inoculation of bartonellae into the blood stream. This phase of disease, termed Oroya fever, is characterized by an acute hemotrophic infection that involves nearly all of the circulating red blood cells. Bartonella's predilection for infecting erythrocytes during this phase is unparalleled among bacterial pathogens. Nearly 80% of the circulating red blood cells are lysed during the attendant hemolytic anemia of Oroya fever, with erythrocyte counts falling from 5 million cells/mm^3 to less than 1 million cells/mm^3.[12,13] It is believed that this radical reduction in hematocrit is caused by splenic culling of the infected red blood cells, although an extracellular β-hemolysin observed by our laboratory might also be involved (Fig. 1).

In addition to severe hemolytic anemia, patients present with chills, head-

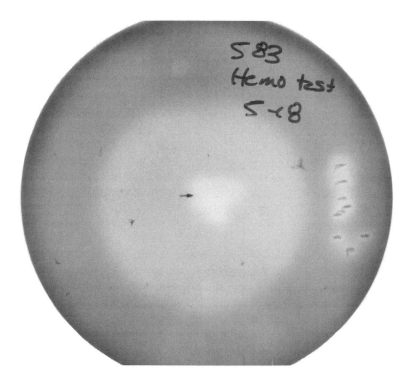

FIGURE 1. Identification of an extracellular β-hemolysin from *B. bacilliformis*. *B. bacilliformis* (strain KC583) was inoculated on the surface of a 0.2-μm filter placed on a heart infusion agar plate containing 2% sheep red blood cells by volume. The filter was removed after 4 days of growth at 30°C. The arrow marks the position of the inoculum and the resulting hemolysis beneath. The large circular area of partial hemolysis indicates the placement site of the filter paper. Control stabs of bartonella can be seen on the right side of the plate and exhibit extensive β-hemolysis. Bacteria were not detectable beneath the filter.

ache, skin pallor, and a febrile response that is possibly enhanced by the lipopolysaccharide (LPS) of the pathogen. The LPS molecule has been partially characterized and is a single diffuse band with an apparent molecular mass of approximately 5 kDa when analyzed by SDS-PAGE.[14,15] The lack of multiple LPS bands on SDS-PAGE suggests that there are only minor variations in the O side chains of the LPS population per given strain of bartonella. Other than its putative role in enhancing fever, the actual involvement of the LPS molecule in virulence is unknown.

Patients who survive the primary phase of bartonellosis generally proceed to the secondary (tissue) phase. At this stage of disease, the pathogen's parasitism shifts from the erythrocytes to the vascular endothelial cells of the capillary bed. Endothelial cells are invaded by the bacterium and many contain extensive intracytosolic vacuoles filled with numerous bacteria, termed rocha lima inclusions. This stage of the disease is generally not fatal and is characterized by the eruption of hemangiomas (soft tumors) on the skin of the extremities and head approximately 4 to 8 weeks following the hematic phase. These hemangiomas are colloquially referred to as verruga peruana and consist of wartlike masses of hyperplastic capillary beds with blood-filled centers containing bartonellae. Hemangiomas can cause cutaneous bleeding, scarring of the patient, and can erupt for up to several months.[8] For unknown reasons, some unusual cases will present with symptoms of the secondary tissue phase without any apparent symptoms of the primary hematic phase. The immunological implications of this are discussed below.

2.2. Colonization

The flagellum is perhaps the most important virulence determinant involved in colonization of the host. *B. bacilliformis* is the only motile member of the order *Rickettsiales*,[16] and each bacterium possesses one to ten polar flagella (Fig. 2). The flagella filaments are composed of multiple 42-kDa flagellin subunits[17] and are markedly resistant to treatment with protease K[18] or trypsin.[17] Flagella have a mean wavelength of approximately 800 nm.[17] The N-terminus of flagellin[17] and the nucleotide sequence of its respective gene (GenBank accession No. L20677) have both been characterized. Flagella, coupled with the normal flow of fluid in the circulatory system, provide bartonella with a high degree of mobility and undoubtedly serve to propel the pathogen during its search for potential host cells. The role of flagella in adhesion and invasion of host cells is discussed below.

2.3. Host Immunity

The human immune response to *B. bacilliformis* produces long-term protective immunity via antibody production, although chronic carriers of the disease are not uncommon.[8] It is still unclear what role the cell-mediated immune response plays in bartonellosis. Although nearly all outer membrane proteins (OMPs) of *B. bacilliformis* can be precipitated during radioimmunoprecipitation experiments employing rabbit hyperimmune antiserum and ^{125}I-labeled bartonella cells, immunogenic OMPs of 31.5, 42, and 45 kDa are the major immunoprecipitants.[15] Our laboratory has tentatively identified two of these proteins as a phage coat protein (31.5 kDa) and the flagellin subunit (42 kDa). The flagellin subunit is possibly an important immunogen for conferring host protection to *B. bacilliformis*. In *in vitro* studies we demonstrated that antiflagellin antibodies were able to reduce *B. bacilliformis* invasion of human erythrocytes by nearly 100% relative to control assays performed with phosphate-buffered saline.[17] In contrast

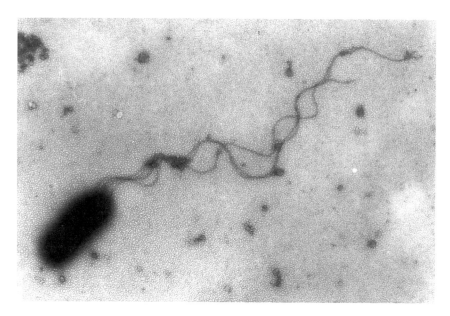

FIGURE 2. Transmission electron micrograph of *B. bacilliformis* (strain KC583). The bacterial cell is approximately 1 μm long with four polar flagella up to 5 μm in length. Bacteria were grown, fixed, and negatively stained with uranyl acetate by the methods of Scherer *et al.*[17]Micrograph is shown at 9000×.

to OMPs, the LPS molecule of *B. bacilliformis* is apparently poorly immunogenic in rabbits,[15] and may generate a poor immune response in humans as well.

Although the immune response to *B. bacilliformis* is poorly characterized, it has been known for some time that humans native to endemic areas are markedly less susceptible to infection and/or the hematic (Oroya fever) phase of disease than are immigrants. Early observations supporting this hypothesis were documented as early as the 16th century by Miguel de Estete and Pizarro's expedition.[19] It is possible that partial or full immune protection in individuals residing in endemic areas may be conferred by neutralizing antibodies generated against recurrent subinfectious exposure(s) to bartonellae or by antibodies obtained through passive maternal acquisition. In previously exposed individuals, a subsequent challenge with an infectious dose results in a nonfatal, i.e., tissue, response. In contrast, travelers to South America who are bitten by *Lutzomyia* are rapidly exposed to an infectious dose of the bacterium via the bloodstream and develop acute Oroya fever.

The correlation between the time course of the two disease phases of bartonellosis and the generation of antibody by the host is intriguing. Weinman noted that verruga production was similar to an immune response, in that it prevented recurrence of acute hematic disease.[8] Interestingly, the acute Oroya fever wanes at about the same time that it takes for significant antibody titers to develop against the pathogen, i.e., about 4 to 6 weeks postinoculation. The formation of verrugas then occurs at 4 to 8 weeks following the hematic phase. It is tempting to suggest that bartonellae are being evicted from the circulatory system by the humoral response and that they subsequently take refuge within the vascular endothelium. However, the observation that viable bartonellae can be obtained from the blood of chronic carriers[6] suggests that bacteria occasionally escape from the endothelium into the bloodstream.

Another immunologic observation associated with Oroya fever is that infected individuals often present with attendant immunosuppression during the hematic phase. The basis for the immunosuppression is unknown, but patients may become susceptible to severe secondary infections with *Salmonella* or *Shigella* or have reactivation of diseases such as tuberculosis or toxoplasmosis.[20,21] Complications produced by secondary infection are believed to enhance the apparent fatality rate of Oroya fever. Fatality rates in untreated cases of Oroya fever range from 40%[9] to 88%.[4] The bacterium is possibly producing a factor(s) that is an active immunosuppressant or is synthesizing a factor that indirectly enhances host modulation of the immune response. The exact nature of the immunosuppressant factor(s) is an interesting problem and remains to be determined.

2.4. Adherence

Studies on the *in vitro* adhesion of bartonella to erythrocytes have shown that there is a 15- to 30-min lag time.[18,22] before bacteria–erythrocyte complexes are formed, and that the maximal number of complexes occurs at about 6 hr.[18] The adhesion process is energy-dependent, as treatment with agents that inactivate bacterial proton-motive force (*N*-ethylmaleimide) or respiration (KCN) significantly reduce the binding of *B. bacilliformis* to red blood cells.[22] By contrast, the erythrocyte is apparently passive in the process and contributes no energy-dependent steps to adhesion. Adherence was not affected by treatment of the erythrocyte with inhibitors of glycolysis (NaF) or proton-motive force (*N*-ethylmaleimide).[22] Alteration of the erythrocyte surface using pronase or subtilisin enhanced adhesion, whereas α- or β-glucosidase treatment decreased adhesion. The authors concluded that protease treatment exposed a glycolipid receptor on the red blood cell surface that could be subsequently destroyed by glucosidase treatment. Although bartonellae are capable of binding erythrocytes from several animal species, it is clear that human erythrocytes are preferentially bound as compared to red blood cells from rabbits or sheep.[22] These observations suggest that human red blood cells may present a more appropriate receptor for the bacterium, or that they possess a greater density and/or a greater accessibility of a receptor that is common to all of the erythrocytes.

The molecular nature of the adhesins responsible for bartonella's adherence to host cells is beginning to unfold. Early work by Walker and Winkler[22] showed that a polar tuft of "fiberlike projections" intimately contacts the erythrocyte surface. The authors noted the resemblance between the adhesive polar tuft and the fibrous meshwork that normally occurs between clumps of bartonella cells observed under microscopy. Recent data from our laboratory show that *B. bacilliformis* possesses aggregative fimbriae. Aggregative fimbriae, also known as bundle-forming pili (BFP), are potentially important adhesin virulence determinants in bacteria such as *Escherichia coli* and *Salmonella enteritidis*.[23] Low-passage *B. bacilliformis* cells that are negatively stained with uranyl acetate and observed by transmission electron microscopy clearly demonstrated randomly distributed fimbriae approximately 5 nm wide on their surface.[24] Using a technique originally designed for enrichment of pili from *Neisseria gonorrhoeae*,[25] we have purified large bundles of putative pili from bartonellae (Fig. 3). The bundles range in size from 50 to 600 nm and are markedly resistant to solubilization with SDS or formic acid.[24]

The flagella may also play a role in the adhesion process. Walker and Winkler[22] suggested that the adhesive polar tuft seen on bartonella might actually be the polar flagella. Benson *et al.*[18] noticed that nonmotile cells bound poorly to red blood cells, but they found no evidence for direct flagellar involvement in adhesion. Their explanation for this observation was that motility might enhance bacteria–erythrocyte collisions or that nonmotile bacteria lacked the adhesin. Similarly, data from our laboratory showed that rabbit antiserum specific to flagellin significantly reduced the *in vitro* association of *B. bacilliformis* with red blood cells by 41%, as compared to controls using preimmune rabbit serum.[17] These data suggest that some, but certainly not all, adherence might be conferred by flagella.

Adhesion to host cells other than erythrocytes has also been investigated. *B. bacilliformis* is capable of *in vitro* association with cultured endothelial [human umbilical vein endothelium (HUVECs)] and epithelial (HEp-2) cells.[26] Association occurs maximally within the first hour of a 3-hr incubation, with equal binding efficiency for both host cell types.

2.5. Invasion

The process by which bartonellae gain entry into host cells likely depends on the type of host cell being invaded. Studies on bartonella invasion of epithelial and endothelial cells show that the host cell is an active participant and can be induced by the pathogen to reconfigure the cytoskeleton to facilitate bacterial uptake.[26] Treatment of HEp-2s or HUVECs with cytochalasin D inhibited invasion by approximately 70%, supporting the hypothesis that actin rearrangement within the host cell is important to invasion. The bacterial component that is

A

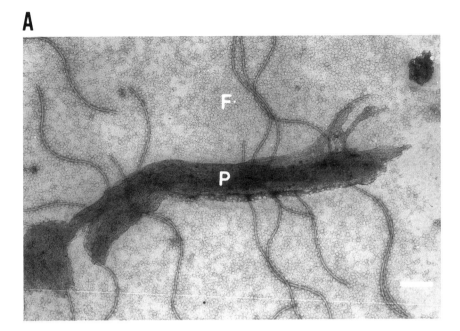

B

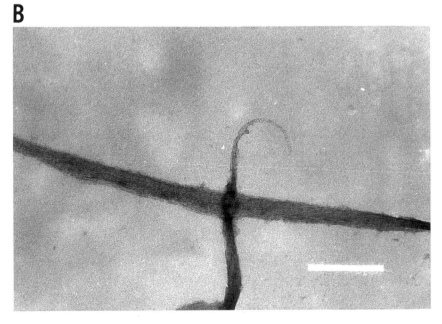

FIGURE 3. Transmission electron micrograph of putative purified pilin bundles from *B. bacilliformis*. Pilin bundles were obtained from strain KC583 by the method of Brinton *et al.*[25] (A) Pilin bundles (P) observed before treatment with 0.1% SDS. Note the numerous contaminating flagella filaments (F) with their periodic curvature and comparatively larger diameter. (B) Purified pilin bundles observed following a 0.1% SDS treatment of the preparation shown in A for 30 mins. A is shown at 20,000×, B at 40,000× (bar = 0.25 μm.)

responsible for inducing microfilament rearrangement in host cells is probably located on the bacterial outer membrane, as antibartonella antiserum significantly reduces uptake of the pathogen by endothelial cell monolayers. The inability to completely inhibit invasion with cytochalasin D suggests that the bacterium is also playing an active role in invasion of epithelial and endothelial cells.[26]

Invasion of human erythrocytes by bartonellae is a very different scenario, as erythrocytes contain very little actin and are essentially nonendocytotic. These characteristics prevent a significant contribution from the erythrocyte during bacterial invasion. Research on the invasion of red blood cells by bartonella implicates three main determinants of virulence, i.e., the deformin protein, flagella, and proteins encoded by the invasion-associated locus (*ialAB*).

Mernaugh and Ihler[27] discovered that *B. bacilliformis* produces an extracellular protein, called deformin, that can produce deep invaginations of erythrocyte membranes independently of the pathogen. Invaginations produced with partially purified deformin are essentially identical to those observed in the infected cells.[18] The deformin protein is 67 kDa on SDS-PAGE and probably occurs as a dimer in its native state.[28] The protein is sensitive to heat or protease treatment and is released into the culture medium during bacterial growth.[27]

The second virulence determinant possibly involved in invasion is the flagellum. Even when deformin is produced, bacteria apparently cannot gain access to the host cell cytosol unless they are motile.[27] To investigate the role of flagella in invasion, our laboratory generated monospecific antibodies against the flagellin subunit and performed invasion assays using antiserum-treated bartonellae. The antiflagellin antibodies nearly abrogated bartonella invasiveness of red blood cells *in vitro*.[17] Taken as a whole, the data suggest that flagella are playing a role in erythrocyte invasion, and together with deformin, flagella may enhance entry of the bacterium into the deformin-produced invaginations.

Recent data from our laboratory also show that *B. bacilliformis* has an invasion locus of approximately 1100 bp, termed *ialAB*.[29] The *ialAB* locus contains two genes (*ialA* and *ialB*) that are both required for invasiveness of transformed *E. coli in vitro*. The *ialA* gene is 510 bp long and encodes a protein that contains a putative NTPase core, whereas the *ialB* gene is 558 bp long and encodes a protein homologue of the Ail (*a*dhesion and *i*nvasion *l*ocus) protein of *Yersinia enterocolitica*.[30] DNA recombinants containing the intact locus can confer an invasive phenotype on minimally invasive strains of *E. coli* (strains HB101 and DH5α) when combined with human red blood cells. *E. coli* containing cloned *ialAB* invaded erythrocytes with a 6- to 39-fold increase relative to controls.[29]

Other than its putative NTPase activity, the function of the IalA protein is unknown. Although still speculative, the IalB protein of *B. bacilliformis* may enhance intracellular survival, much like Ail does for yersiniae.[31] The most exciting aspects of the *ialAB* locus are that (1) a single bartonella locus can confer an invasive phenotype on *E. coli*, (2) the invasive phenotype requires both *ialA* and

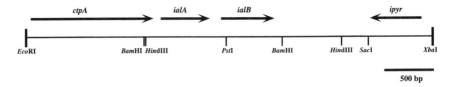

FIGURE 4. Partial restriction and linkage map of the invasion-associated locus (*ialAB*) of *B. bacilliformis*[29] and its flanking sequences. Open reading frames are shown as large arrows with their respective gene designations above, including *ctpA* (C-terminal processing protease), *ialA* (invasion-associated locus A), *ialB* (invasion-associated locus B), and *ipyr* (inorganic pyrophosphatase). Restriction endonuclease sites are indicated below the map.

ialB genes to be present, and (3) one of the genes, *ialB*, is a homologue to a known invasion-associated bacterial gene product, the Ail protein of *Yersinia*. The actual mechanism by which *ialAB* mediates erythrocyte invasion and whether the locus mediates or facilitates the invasion of other cell types remains to be determined. We are presently characterizing the flanking sequences of the *ialAB* locus to determine if other virulence determinants are closely linked. To date, we have discovered a C-terminal protease gene (*ctpA*; GenBank accession No. L37094) that lies 69 bp 5′ to the *ialAB* locus and an inorganic pyrophosphatase gene (*ipyrA*; GenBank accession No. L46591) that occurs 1022 bp 3′ to the locus (Fig. 4). The role of *ipyrA* and *ctpA* in virulence is not known but is currently under investigation in our laboratory.

Electron microscopy of *ialAB*-transformed *E. coli* invading erythrocytes reveals a similar picture to what is observed when bartonellae invade erythrocytes. Complementary invagination occurs in the erythrocyte membrane around each invading bacterium and is subsequently followed by vacuole formation within the cytosol (Fig. 5). The vacuoles containing *E. coli* do not appear to be membrane-bound and have a large space surrounding the bacterium, as has also been observed by electron microscopy of bartonella invasion of red blood cells.[34] These observations are in contrast to those seen with light microscopy, which show a distinct membrane-bound vacuole that can retain fluorescein isothiocyanate and exclude calcein dyes.[18] It is possible that fixing of the red blood cells prior to electron microscopy may enhance the vacuolar space around the bacterium and somehow mask the unit membrane.

3. SURVIVAL AND REPLICATION WITHIN THE HOST

3.1. The Erythrocyte

B. bacilliformis can efficiently enter a variety of host cells, but the underlying purpose(s) for invasion is still debatable. The hemotrophic nature of the bacte-

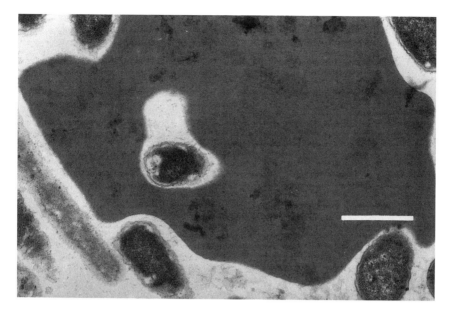

FIGURE 5. Transmission electron micrograph of *E. coli* (pIAL1) located within a human erythrocyte vacuole following invasion. *E. coli* (strain HB101) was transformed with a recombinant containing the *ialAB* locus of *B. bacilliformis* (pIAL1). Invasion assays were conducted by a modification of the gentamicin protection assay of Isberg *et al.*[32,33] Micrograph is shown at 10,000× (bar = 1 μm).

rium suggests that the pathogen requires a blood-supplied growth factor(s) such as iron, and indeed the bacterium requires blood for any growth to occur *in vitro*.[16] Invasion or hemolysis of erythrocytes would provide the bacterium with an ample supply of growth factor(s) such as iron. In addition, the intracellular environment provides an ideal location for evading components of the host's humoral immune response.

Entry into erythrocytes and formation of the erythrocyte endosomal vacuole is still unclear. The main question that begs to be answered is how the bacterium enters a red blood cell without lysing it. The cytosolic membrane of the red blood cell must rapidly fuse at the site of bacterial penetration to prevent immediate hemolysis. Early investigators observed complementary deformation of e erythrocyte membrane around bartonellae and theorized that flagella-mediated propulsion was responsible for the invaginations.[18] Although this may occur, the discovery of an invasion locus and deformin suggest that flagella may actually play a secondary role in the invasion process, perhaps by propelling the pathogen into deformin-produced trenches and pits. Once inside the invaginations the bacterium probably attaches and employs IalA and IalB to actually penetrate the red blood cell membrane. The mechanics of IalAB-mediated invasion is presently under investigation in our laboratory. The flagella are probably not necessary to complete the invasion process, as nonflagellated strains of *E. coli* (i.e., HB101 and DH5α) that contain the cloned *ialAB* locus of *B. bacilliformis* can enter human erythrocytes.[29] However, flagella are apparently retained by bartonella after invading the host cell,[18] implying a continued need for their service.

Bartonellae have also been observed to replicate within, and occasionally escape from, the vacuoles of erythrocytes.[18] The rupture of these vacuoles probably does not result from overcrowding with replicating bacteria, as vacuoles can markedly enlarge over time so as to accommodate new bacterial cells.[18] Because flagella are retained by bartonella within endosomal vacuoles, it is possible that they are employed by the pathogen to escape from vacuoles or the cytoplasmic membrane of the host cell. It is also possible that the hemolytic activity of bartonellae observed on semisolid medium containing blood (Fig. 1) weakens the host cell membranes of the vacuole or the cytosol. The concerted effort of hemolysin(s) and flagella could facilitate invasion or escape from the vacuole or host cell cytoplasmic membrane.

3.2. Endothelial Cells

During the secondary (tissue) phase of disease, *B. bacilliformis* shows a marked predilection for infecting the endothelial cells of the capillary beds of the skin and subcutaneous tissue. Hemangioma formation involving deep organs has not been documented and may reflect a temperature preference by the pathogen.[35] At least two distinct histological manifestations of verruga peruana are known.[35] A spindly form contains an uncharacterized fusiform cell with possible T cell and monocyte/macrophage infiltration, whereas a pyogenic granulomatous hemangioma is characterized by hyperplastic endothelium and extensive vascularization. Histologically, the spindle-form lesions are similar to Kaposi's sarcoma,[36] whereas the granulomatous form resembles the cutaneous lesions of bacillary angiomatosis seen during infection with *B. henselae*.[37] *In vitro* work by Garcia *et al.*[35] shows that viable bartonellae can induce HUVEC proliferation and rocha lima inclusions. The verruga usually contain numerous bartonellae, epithelioid macrophage cells, and several inflammatory cells such as polymorphonuclear cells.

B. bacilliformis produces an angiogenic factor that is both mitogenic for endothelial cells and can cause angiogenesis (vascularization) in an *in vivo* rat model. Garcia *et al.*[38] showed that a mitogenic fraction could be collected from extracts of bartonella grown *in vitro*. The mitogenic component is probably proteinaceous, as it was heat sensitive, possessed a molecular mass greater than 12 to 14 kDa, and was precipitated with 45% ammonium sulfate. The factor's activity is not enhanced with heparin, nor does the factor bind heparin. *In vitro*, HUVEC growth was increased approximately 3-fold over controls stimulated with *Haemophilus influenzae* extracts. In addition, mitogenicity was specific to endothelial cells;

smooth muscle and mesenchyme growth was unaffected. The authors also showed that the bartonella extract-treated HUVECs were releasing a marker protein for angiogenesis, i.e., tissue-type plasminogen activator (t-PA), in a dose-dependent fashion. *In vivo*, the bartonella extract enhanced the formation of blood vessels by 2.5-fold in rats with subcutaneous sponge implants containing the extract. The combination of t-PA production *in vitro* and the enhanced formation of blood vessels *in vivo* suggests that the angiogenic factor of bartonella is genuine. However, it is possible that the factor is indirectly angiogenic; relying on host-derived factors to induce the actual angiogenic response. The molecular nature of the angiogenic factor(s) is an intriguing problem and remains to be determined.

4. CONCLUSIONS

Understanding the molecular basis for pathogenesis in *B. bacilliformis* will serve as a model system to study virulence mechanisms in other hemotrophic bacterial pathogens. A more comprehensive understanding of pathogenesis is important, as many species of bartonella, including *B. quintana*, *B. henselae*, and *B. elizabethae*, are recognized as emerging pathogens of clinical importance. Examples that illustrate the increasing clinical impact of the genus include the secondary infection of AIDS patients with *B. henselae* and the recent outbreak of urban trench fever (*B. quintana*) in Seattle.[39] The development of a transformation system using electroporation[40] and the genetic mapping of the *B. bacilliformis* genome[41] are enhancing our ability to continue the characterization and genetic manipulation of the virulence factors of *B. bacilliformis*.

ACKNOWLEDGMENTS. I thank my past and present graduate students David Scherer, Helen Grasseschi, Samuel Mitchell, Steven McAllister, and James Battisti for their excellent scientific contributions to this area of science. I gratefully acknowledge the financial support of our research by Public Health Service grants AI34050 and RR10169 from the National Institutes of Health.

REFERENCES

1. Schultz, M. G., 1968, A history of bartonellosis (Carrion's disease), *Am. J. Trop. Med. Hyg.* **17:**503–515.
2. Matteelli, A., Castelli, F., Spinetti, A., Bonetti, F., Graifenberghi, S., and Carosi, G., 1994, Short report: Verruga peruana in an Italian traveler from Peru, *Am. J. Trop. Med. Hyg.* **50:**143–144.
3. Alexander, B., 1995, A review of bartonellosis in Ecuador and Colombia, *Am. J. Trop. Med. Hyg.* **52:**354–359.
4. Gray, G. C., Johnson, A. A., Thornton, S. A., Smith, W. A., Knobloch, J., Kelley, P. W., Escudero, L. O., Huayda, M. A., and Wignall, F. S., 1990, An epidemic of Oroya fever in the Peruvian Andes, *Am. T. Trop. Med. Hyg.* **42:**215–221.

5. Knobloch, J., Solano, L., Alvarez, O., and Delgado, E., 1985, Antibodies to *Bartonella bacilliformis* as determined by fluorescence antibody test, indirect hemagglutination and ELISA, *Trop. Med. Parasitol.* **36**:183–185.

6. Herrer, A., 1953, Presence of *Bartonella bacilliformis* in the peripheral blood of patients with the benign form, *Am. J. Trop. Med.* **2**:645–649.

7. Strong, R. P., 1918, Trench fever: Report of commission, medical research committee, American Red Cross, Oxford University Press, pp. 40–60.

8. Weinman, D., 1965, The bartonella group, in: *Bacterial and Mycotic Infections of Man* (R. J. Dubos and J. G. Hirsch, eds.), Lippincott, Philadelphia, pp. 775–785.

9. Weinman, D., 1944, Infectious anemias due to bartonella and related red cell parasites, *Trans. Am. Philos. Soc.* **33**:243–287.

10. Kreier, J. P., and Ristic, M., 1981, The biology of hemotrophic bacteria, *Annu. Rev. Microbiol.* **35**:325–338.

11. Noguchi, H., 1926, The etiology of Oroya fever, *J. Exp. Med.* **43**:851–864.

12. Hurtado, A., Musso, J. P., and Meriono, C., 1938, La anemia en la enfermedad de Carrion (verruga peruana), *Ann. Fac. Med. Lima* **28**:154–168.

13. Reynafarje, C., and Ramos, J., 1961, The hemolytic anemia of human bartonellosis, *Blood* **17**:562–578.

14. Knobloch, J., Bialek, R., Muller, G., and Asmus, P., 1988, Common surface epitope of *Bartonella bacilliformis* and *Chlamydia psittaci*, *Am. J. Trop. Med. Hyg.* **39**:427–433.

15. Minnick, M. F., 1994, Identification of outer membrane proteins of *Bartonella bacilliformis*, *Infect. Immun.* **62**:2644–2648.

16. Ristic, R., and Kreier, J. P., 1984, Family II. *Bartonellaceae* Gieszczykiewicz 1939, 25AL, in: *Bergey's Manual of Systematic Bacteriology*, Volume 1 (N. R. Krieg and J. G. Holt, eds.), Williams & Wilkins, Baltimore, pp. 717–719.

17. Scherer, D. C., DeBuron-Connors, I., and Minnick, M. F., 1993, Characterization of *Bartonella bacilliformis* flagella and effect of antiflagellin antibodies on invasion of human erythrocytes, *Infect. Immun.* **61**:4962–4971.

18. Benson, L. A., Kar, S., McLaughlin, G., and Ihler, G. M., 1986, Entry of *Bartonella bacilliformis* into erythrocytes, *Infect. Immun.* **54**:347–353.

19. Strong, R. P., Tyzzer, E. E., Brues, C. T., Sellards, A. W., and Gastiaburu, J. C., 1915, in: *Report of First Expedition to South America, 1913*, Harvard School of Tropical Medicine, Harvard University Press, Cambridge, MA, pp. 3–175.

20. Cuadra, C. M., 1956, Salmonellosis complication in human bartonellosis, *Tex. Rep. Biol. Med.* **14**:97–113.

21. Garcia-Caceres, U., and Garcia, F. U., 1991, Bartonellosis: An immunodepressive disease and the life of Daniel Alcides Carrión, *Am. J. Clin. Pathol.* **95**:S58–S66.

22. Walker, T. S., and Winkler, H. H., 1981, *Bartonella bacilliformis*: Colonial types and erythrocyte adherence, *Infect. Immun.* **31**:480–486.

23. Sjobring, U., Pohl, G., and Olsen, A., 1994, Plasminogen, absorbed by *Escherichia coli* expressing curli or by *Salmonella enteritidis* expressing thin aggregative fimbriae, can be activated by simultaneously captured tissue-type plasminogen activator (t-PA), *Mol. Microbiol.* **14**:443–452.

24. McAllister, S. J., Peek, J. A., and Minnick, M. F., 1995, Identification and isolation of bundle-forming fimbriae from *Bartonella bacilliformis*, in: *Abstracts of the 95th General Meeting of the American Society for Microbiology*, American Society for Microbiology, Washington, DC, abstr. D-43, p. 256.

25. Brinton, C. C., Bryan, J., Dillon, J.-A., Guerina, N., Jacobson, L. J., Labik, A., Lee, S., Levine, A., Lim, S., McMichael, J., Polen, S., Rogers, K., To, A., C.-C., and C.-C. To, S. C.-M., 1978, Uses of pili in gonorrhea control: Role of bacterial pili in disease, purification and properties of gonococcal pili, and progress in the development of a gonococcal pilus vaccine for gonorrhea. in:

Immunobiology of Neisseria gonorrhoeae (G. F. Brooks, E. C. Gotschlich, K. K. Holmes, W. D. Sawyer, and F. E. Young, eds.), American Society for Microbiology, Washington, DC, pp. 18–20.

26. McGinnis-Hill, E., Raji, A., Valenzuela, M. S., Garcia, F., and Hoover, R., 1992, Adhesion to and invasion of cultured human cells by *Bartonella bacilliformis*, *Infect. Immun.* **60:**4051–4058.

27. Mernaugh, G., and Ihler, G. M., 1992, Deformation factor: An extracellular protein synthesized by *Bartonella bacilliformis* that deforms erythrocyte membranes, *Infect. Immun.* **60:**937–943.

28. Xu, Y.-H., Lu, Z.-Y., and Ihler, G. M., 1995, Purification of deformin, an extracellular protein synthesized by *Bartonella bacilliformis* which causes deformation of erythrocyte membranes, *Biochim. Biophys. Acta* **1234:**173–183.

29. Mitchell, S. J., and Minnick, M. F., 1995, Characterization of a two-gene locus from *Bartonella bacilliformis* associated with the ability to invade human erythrocytes, *Infect. Immun.* **63:**1552–1562.

30. Miller, V. L., and Falkow, S., 1988, Evidence for two genetic loci in *Yersinia enterocolitica* that can promote invasion of epithelial cells, *Infect. Immun.* **56:**1242–1248.

31. Miller, V. L., Bliska, J. B., and Falkow, S., 1990, Nucleotide sequence of the *Yersinia enterocolitica ail* gene and characterization of the Ail protein product, *J. Bacteriol.* **172:**1062–1069.

32. Isberg, R. R., and Falkow, S., 1985, A single genetic locus encoded by *Yersinia pseudotuberculosis* permits invasion of cultured animal cells by *E. coli* K12, *Nature* **317:**262–264.

33. Isberg, R. R., Voorhis, D. L., and Falkow, S., 1987, Identification of invasin: A protein that allows enteric bacteria to penetrate cultured mammalian cells, *Cell* **50:**769–778.

34. Cuadra, M., and Takano, J., 1969, The relationship of *Bartonella bacilliformis* to the red blood cell as revealed by electron microscopy, *Blood* **33:**708–716.

35. Garcia, F. U., Wojta, J., and Hoover, R. L., 1992, Interactions between live *Bartonella bacilliformis* and endothelial cells, *J. Infect. Dis.* **165:**1138–1141.

36. Arias-Stella, J., Lieberman, P. H., Erlandson, R. A., and Arias-Stella, J., Jr., 1986, Histology, immunochemistry and ultrastructure of the verruga in Carrion disease, *Am. J. Surg. Pathol.* **10:**595–610.

37. LeBoit, P. E., Berger, T. G., Egbert, B. M., Beckstead, J. H., Benedict-Yen, T. S., and Stoler, M. H., 1989, Bacillary angiomatosis. The histopathology and differential diagnosis of a pseudo-neoplastic infection in patients with human immunodeficiency virus, *Am. J. Surg. Pathol.* **13:**909–920.

38. Garcia, F. U., Wojta, J., Broadley, K. N., Davidson, J. M., and Hoover, R. L., 1990, *Bartonella bacilliformis* stimulates endothelial cells *in vitro* and is angiogenic *in vivo*, *Am. J. Pathol.* **136:**1125–1135.

39. Regnery, R., and Tappero, J., 1995, Unraveling mysteries associated with cat-scratch disease, bacillary angiomatosis, and related syndromes, *Emerg. Infect. Dis.* **1:**16 –21.

40. Grasseschi, H. A., and Minnick, M. F., 1994, Transformation of *Bartonella bacilliformis* by electroporation, *Can. J. Microbiol.* **40:**782–786.

41. Krueger, C. M., Marks, K. L., and Ihler, G. M., 1995, Physical map of the *Bartonella bacilliformis* genome, *J. Bacteriol.* **177:**7271–7274.

13

Stimulation of Angiogenesis and Protection from Oxidant Damage
Two Potential Mechanisms Involved in Pathogenesis by *Bartonella henselae* and Other *Bartonella* Species

THOMAS D. CONLEY, MATTHEW F. WACK,
KAREN K. HAMILTON, and LEONARD N. SLATER

1. INTRODUCTION

Our understanding of *Bartonella* pathogenesis remains limited despite long-standing recognition of clinical syndromes. Human illness attributable to *B. bacilliformis* and *B. quintana* has been recognized for decades. However, investigations directed at their pathogenic effects have been limited, perhaps because of the geographic limitation of disease caused by the former and the sporadic nature of disease

THOMAS D. CONLEY and KAREN K. HAMILTON • Cardiology Section, The University of Oklahoma Health Sciences Center, Oklahoma City, Oklahoma 73104.   MATTHEW F. WACK and LEONARD N. SLATER • Infectious Diseases Section, The University of Oklahoma Health Sciences Center, and The Department of Veterans Affairs Medical Center, Oklahoma City, Oklahoma 73104. *Present address of T.D.C.:* University of Arkansas for Medical Science, Little Rock, Arkansas 72204. *Present address of K.K.H.:* Cardiology Associates of Corpus Christi, Corpus Christi, Texas 78412. *Present address of M.F.W.:* Infectious Disease of Indiana, Indianapolis, Indiana 46202.

Rickettsial Infection and Immunity, edited by Anderson *et al.* Plenum Press, New York, 1997.

caused by the latter. Renewed interest in *B. bacilliformis* has developed during the past two decades. The past half decade has seen the identification of new species, including *B. henselae* and *B. elizabethae,* and the implication of *B. henselae* in cat-scratch disease, of *B. henselae* and *B. quintana* in bacillary angiomatosis/peliosis, and of *B. henselae, B. quintana,* and *B. elizabethae* in bacteremia with endocarditis. Although most pathogenesis work to date has been accomplished using *B. bacilliformis,* information on pathogenic mechanisms of other species is now becoming available.

Two facets of pathogenesis, potentially interrelated, have been the focus of our investigations: stimulation of angiogenesis and bacterial antioxidant defenses. Each area shall be reviewed in this chapter.

2. STIMULATION OF ANGIOGENESIS

2.1. Background

Angiogenesis is the development of new blood vessels from preexisting vessels. *In vivo,* this is a complex process in which endothelial cells and pericytes from preexisting capillaries, under the influence of angiogenic growth factors, migrate, proliferate, and differentiate to form new blood vessels. This process begins with the loss of the basal lamina of small vessels, which is temporally related to endothelial cell production of tissue plasminogen activator (t-PA) and collagenases capable of enzymatic degradation of the basement membrane. Subsequently, endothelial cells extend cytoplasmic pseudopodia into the interstitium toward a stimulus, followed by migration of the endothelial cells and then of pericytes. Differentiation of these migrating cells results in the formation of new vascular channels.[1]

Stimulated directional migration of capillary endothelial cells appears to dominate subsequent events in capillary growth and occurs in response to angiogenic growth factors. These substances, usually proteins, exert their effects on certain target cells (including endothelial and vascular smooth muscle cells) and thereby modulate angiogenesis. Several angiogenic growth factors have been characterized biochemically and their mitogenic and differentiation effects on endothelial cells described. These factors also exhibit other angiogenesis-related biological activities including chemotactic and/or mitogenic stimulation for a variety of cell types, heparin-binding, and stimulation of DNA synthesis, protein secretion, and extracellular matrix production.[2,3]

Several models have been developed for the study of angiogenesis. The endothelial cell response to potential growth factors is frequently used as an *in vitro* model because of the cardinal roles the endothelial cell plays in new blood vessels. Frequently measured endothelial cell responses include production of factors that cause enzymatic degradation of basement membrane (e.g., t-PA), proliferation, chemotaxis, changes in cellular cytoskeleton, and differentiation (e.g., formation of tubules in a collagen matrix). There are several *in vivo* models available to study angiogenic substances. These models assay new blood vessel formation after application of potential angiogenic agents into the normally avascular rabbit cornea, into sponges implanted in the subcutaneous tissues of rodents, or onto the chorioallantoic membrane of a chicken embryo.

Important in fetal development and normal growth, angiogenesis ordinarily is a strictly controlled process that occurs only under selected circumstances in fully grown humans. For example, the formation of new blood vessels is essential for effective wound healing, cyclical endometrial proliferation, muscle hypertrophy in response to exercise, and collateral blood vessel formation. Certain pathologic states such as diabetic retinopathy and tumor growth are characterized by the formation of new blood vessels as well. Unique among infectious disorders, the syndromes of verruga peruana and bacillary angiomatosis/peliosis associated with *Bartonella* infections are also characterized by aberrant neoangiogenesis.[4-7] As these phenomena regress with antimicrobial therapy, it is reasonable to suspect that bacterial growth/metabolism plays a role in their development and persistence in the absence of antibiotics. Indeed, availabled data suggest that these syndromes may be caused by angiogenic growth factors produced by the bacteria.

2.2. *B. bacilliformis* Stimulation of Endothelial Cells

In vitro evidence of bacterial stimulation of angiogenesis was first developed by Garcia *et al.*[8] using a crude extract of *B. bacilliformis* produced through sonic disruption and centrifugation. When human umbilical vein endothelial cells (HUVEC) or peripheral vein endothelial cells were exposed to growth media containing selected concentrations of this extract, endothelial cell growth, as determined by counting cells, was enhanced as much as 300% of control cells. Such stimulatory effects of the extract were not seen with smooth muscle or mesenchymal cells. The extract also stimulated endothelial cell production of t-PA. Both endothelial cell proliferation and t-PA production were concentration dependent; t-PA production was time dependent as well. Further analysis demonstrated that the stimulatory factor contained within the extract was heat sensitive, precipitable by 45% ammonium sulfate, and larger than 12–14 kDa (as determined by dialysis). It neither was enhanced by nor had an affinity for heparin. The crude extract also stimulated angiogenesis *in vivo*. Introduced subcutaneously into rats via impregnated sponges, it caused new blood vessel formation within 4 days (Fig. 1).

Subsequent studies by Garcia *et al.*[9] investigated the relationship between intact *B. bacilliformis* and HUVEC. When endothelial cells were coincubated with live *B. bacilliformis*, the bacteria penetrated endothelial cells within an hour and by

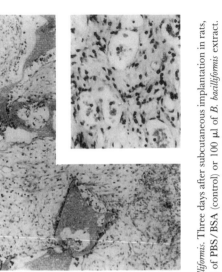

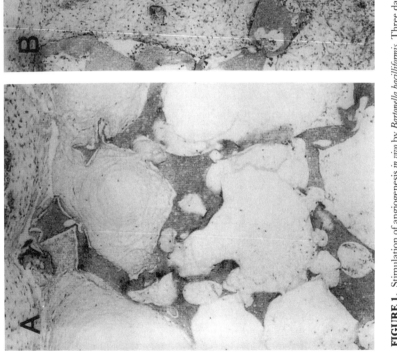

FIGURE 1. Stimulation of angiogenesis *in vivo* by *Bartonella bacilliformis*. Three days after subcutaneous implantation in rats, polyvinyl alcohol sponge disks were injected with either 100 μl of PBS/BSA (control) or 100 μl of *B. bacilliformis* extract. Local histology 4 days after the disk injections was investigated by light microscopic review of formalin-fixed, hematoxylin and eosin-stained sections. Panel A demonstrates the lack of cellular infiltration of a control disk (original magnification 80×). Panel B demonstrates granulation tissue infiltrating a disk injected with the extract (original magnification 80×), and the inset demonstrates newly formed blood vessels but no inflammatory cells within the granulation tissue (original magnification 180×). (Reproduced from Garcia *et al.*[8])

12 h formed large, membrane-bound inclusions (Fig. 2). The investigators suggested these resembled the Rocha-Lima inclusions seen in verruga peruana. (However, modern ultrastructural studies have cast doubt on the contention that such inclusions represent bacterial clusters.[4]) Nevertheless, coincubation of *B. bacilliformis* and HUVEC did result in increased endothelial cell proliferation. Such proliferation was evident when bacteria and endothelial cells were physically separated by a membrane, but was greater when bacteria and HUVEC were admixed during coincubation. These findings indicated that *B. bacilliformis* could shed angiogenic factor into the local environment, but also suggested that other bacterial–endothelial cell interactions played a role in stimulating HUVEC proliferation. Incubation of live *B. bacilliformis* with endothelial cells also resulted in the production of t-PA, similar to the effect produced by the crude extract. Taken together, these studies provided the first evidence that *B. bacilliformis* could induce angiogenesis in a primary fashion, likely through its production of a factor that stimulated endothelial cell growth.

2.3. *B. henselae* and *quintana* Stimulation of Endothelial Cells

Contemporaneous with the reports by Garcia and collaborators, data emerged implicating *Bartonella* (then *Rochalimaea*) *henselae* and *quintana* in the pathogenesis of bacillary angiomatosis and peliosis.[10–15] Given the clinical and histologic similarities between bacillary angiomatosis and verruga peruana, it was reasonable to suspect that *B. henselae* and *B. quintana* also should possess angiogenic activity. Although Koehler's first report of culture isolation of these agents from lesions of bacillary angiomatosis utilized a system of coculture with bovine endothelial cells,[15] it did not detail whether endothelial cell growth appeared stimulated. Conley *et al.*[16] since have reported some of the effects of *B. henselae* and other *Bartonella* species on human endothelial cell growth *in vitro*. *B. henselae* could be cocultivated with HUVEC and viable bacteria recovered from the endothelial cell monolayers for several weeks. When cultured in the presence of viable *B. henselae,* endothelial cell growth was markedly enhanced in a period of days compared to control cells (Fig. 3). Such effects subsequently were demonstrated with multiple clinical isolates of *B. henselae* and *B. quintana* (Fig. 4). HUVEC stimulation increased with the concentration of bacteria applied; similar studies conducted with fibroblasts demonstrated no effect on growth (Fig. 5). In addition to stimulating *in vitro* cellular proliferation, live *Bartonella* spp. stimulated another aspect of angiogenesis, endothelial cell migration (Fig. 6).

These effects were not seen exclusively with live organisms. When crude fractions were created by sonic disruption of bacterial cells and centrifugation, the ability to stimulate endothelial cell proliferation and migration resided in a particulate, noncytosolic fraction (Fig. 7). The stimulatory effect of this largely undefined fraction was markedly diminished by exposure to trypsin, suggesting that

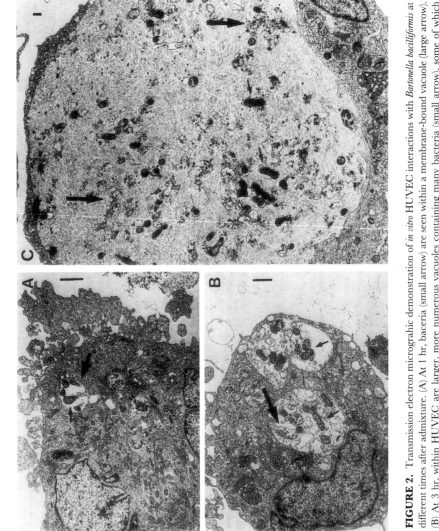

FIGURE 2. Transmission electron micrograhic demonstration of *in vitro* HUVEC interactions with *Bartonella bacilliformis* at different times after admixture. (A) At 1 hr, bacteria (small arrow) are seen within a membrane-bound vacuole (large arrow). (B) At 3 hr, within HUVEC are larger, more numerous vacuoles containing many bacteria (small arrow), some of which appear to be undergoing degradation (large arrow). (C) At 12 hr, a single large vacuole occupying most of the HUVEC cytoplasm, perhaps representing fusion of smaller vacuoles, contains both apparently intact bacteria (small arrows) and amorphous material (large arrow) which may represent degraded bacteria. Bars = 1 μm. (Reproduced from Garcia *et al.*[9])

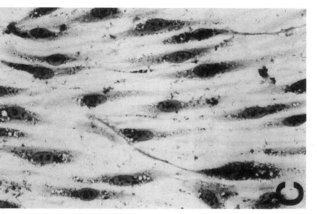

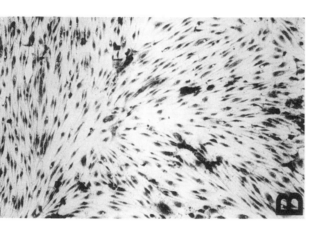

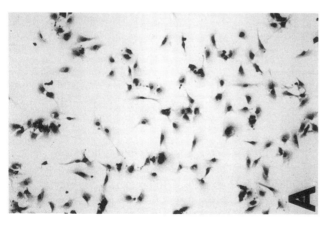

FIGURE 3. Photomicrographs of human umbilical vein endothelial cells cocultivated with *B. henselae*. HUVEC were grown in multiwell microscope slides containing either Medium 199 + 20% FCS (control, panel A) or live *B. henselae* [strain 87-66 (= American Type Culture Collection No. 49793) at a protein concentration of 33 μg/ml] in the same culture medium (panels B and C). Cells were fed with culture medium every 48 hr, then stained and photographed after 6 days (Wright–Giemsa stain, original magnification 100× for panels A and B, 1000× for panel C). Panel B demonstrates the increase in density and morphologic change of HUVEC, relative to the control in panel A, resulting from coincubation with *B. henselae*. Panel C demonstrates the presence of apparently adherent bacteria and marked vacuolation of the HUVEC.

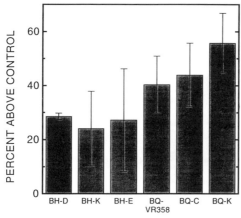

FIGURE 4. The effect of *Bartonella* spp. on HUVEC proliferation. Selected isolates of *Bartonella henselae* (BH), and *Bartonella quintana* (BQ) were evaluated for their potential to stimulate HUVEC growth. BH-D (strain 87-66, ATCC 49793) was isolated from an HIV-infected man with bacteremia and bacillary peliosis; BH-K was isolated from an immunocompetent man with bacteremia; BH-E was isolated from an immunosuppressed renal transplant recipient with bacillary angiomatosis and peliosis. BQ-VR358 is the "Fuller" type strain. BQ-C was isolated from a patient with AIDS and bacteremia. BQ-K was isolated from an HIV-infected patient with endocarditis.

HUVEC were seeded in 96-well plates containing Medium 199 with 20% FCS. After 4 hr, selected concentrations of *Bartonella* isolates (range 0.3 to 10 μg protein/ml) were added. Six days later, proliferation was assessed using an acid phosphatase assay. The results shown represent the maximal amount of stimulation produced by each isolate over the range of concentrations tested. The data are expressed as percent above control ($n = 6$, mean ± SD). (Adapted from Conley et al.[16])

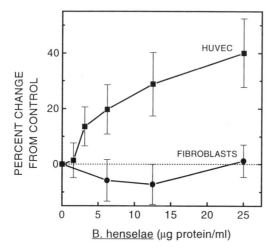

FIGURE 5. The relative effects of *B. henselae* on HUVEC and fibroblast proliferation. HUVEC or fibroblasts were seeded in 96-well plates containing either Medium 199 with 20% FCS or MEM with 10% FCS, respectively. After 4 hr, selected concentrations of *B. henselae* (strain 87-66, 0 to 25 μg protein/ml) were added. Cell proliferation was assessed by the detection of acid phosphatase activity after 6 days. Results are expressed as percent change from control (HUVEC $n = 7$; fibroblasts $n = 10$; mean ± SD). (Adapted from Conley et al.[16])

the angiogenic factor is a protein (Fig. 8). Further purification of a *B. henselae* membrane preparation by detergent extraction, separation with ion exchange chromatography, and dialysis through 12–14,000 molecular weight membranes has resulted in the recovery of an angiogenic factor capable of producing an approximately 50% increase in HUVEC growth (unpublished data).

FIGURE 6. The effect of *Bartonella* spp. on HUVEC migration. BH-S was isolated from a patient with bacteremia whose immunologic status is unknown. *B. bacilliformis* was obtained from the ATCC (No. 35685). Other isolates are as described in Fig. 4. HUVEC were seeded onto 3-μm-pore polycarbonate membranes in modified Boyden chambers containing either medium only (CONTROL) or suspensions of live bacteria (10 μg protein/ml medium). After 18 hr, the membranes were washed with HBSS.

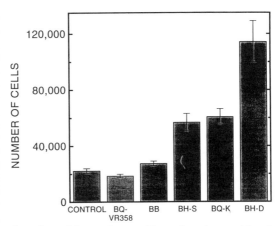

HUVEC were removed from the undersurfaces of the membranes with trypsin and counted (n = 3, mean ± SD). (Adapted from Conley *et al.*[16])

In summary, human infection by *B. bacilliformis*, *B. henselae*, and *B. quintana* can result in the development of neovascular lesions which characteristically regress with antibiotic treatment. In *in vitro* and *in vivo* experimental models, these species can induce both the proliferation and migration of endothelial cells, likely through the production of protein(s) localized in the bacterial cell wall/membrane region having angiogenic growth factor characteristics. The production of an angiogenic growth factor by *Bartonella* spp. is apparently unique among prokaryotes. Speculation about the survival/replication advantages of such a factor is tempting, but the data are as yet too few. Further studies are needed to develop a more complete understanding of this unusual pathogenic feature.

3. PROTECTION FROM OXIDANT DAMAGE

3.1. Background

The incomplete reduction of molecular oxygen leads to the formation of several potentially toxic products. These include superoxide radical ($\cdot O_2^-$) which is formed directly from the partial reduction of molecular oxygen; hydrogen peroxide which is formed by the dismutation of superoxide anion ($2 \cdot O_2^- + 2H^+ \rightarrow H_2O_2 + O_2$); and possibly hydroxyl radical ($\cdot OH$) generated by Fe-dependent catalytic conversion of H_2O_2, as predicted by the Fenton reaction ($H_2O_2 + Fe^{2+} \rightarrow \cdot OH + {}^-OH + Fe^{3+}$).[17,18] Exposure to these radical species can result in cytotoxic events, such as DNA damage, oxidation of critical proteins, and lipid peroxidation resulting in loss of membrane integrity. These oxygen radicals are

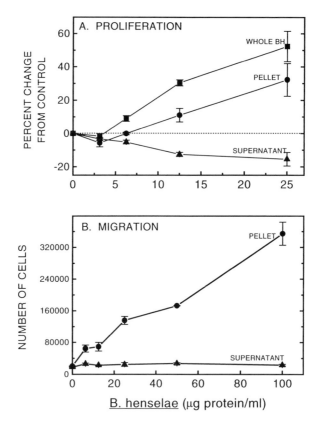

FIGURE 7. The effect of sonic disruption and centrifugation on the ability of *B. henselae* to stimulate HUVEC proliferation and migraiton. A suspension of *B. henselae* strain 87-66 was subjected to sonic disruption while in an ice bath, then the resulting lysate centrifuged at 3000*g* for 3 min. The initial pellet containing large bacterial fragments was discarded and the supernatant then centrifuged at 13,000*g* for 30 min. The final supernatant was decanted and the resulting final pellet resuspended in sterile HBSS.

(A) HUVEC growth in response to selected concentrations of whole *B. henselae*, supernatant, or pellet (0 to 25 μg protein/ml) was assessed after 6 days by the detection of acid phosphatase activity (*n* = 6, mean ± SD). The data are expressed as percent change from control (no bacterial protein).

(B) HUVEC migration in response to *B. henselae* pellet or supernatant (0 to 100 μg protein/ml) was assessed after 18 hr as described in Fig. 6 (*n* = 3, mean ± SD). (Adapted from Conley *et al.*[16])

generated during metabolic events involving molecular oxygen, such as respiration, the degradation of oxyhemoglobin,[19] and the respiratory burst of stimulated phagocytes.[20]

Pathogenic *Bartonella* species survive in a variety of potentially oxidative environments during infection of the host. Therefore, the ability of *Bartonella* spp.

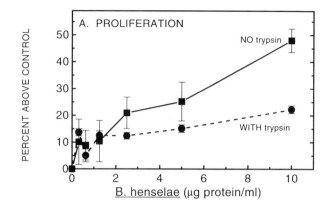

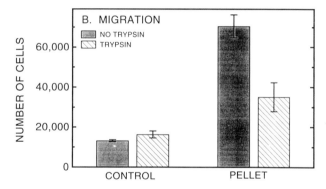

FIGURE 8. The effect of trypsin on the ability of *B. henselae* to stimulate HUVEC migration and proliferation. Resuspended *B. henselae* strain 87-66 final pellet fraction was prepared as described in Fig. 7. Pellet suspension and buffer controls were incubated in the presence or absence of trypsin (0.83 mg/ml) at 37°C for 15 min, then combined with FCS to inactivate the trypsin. Protein concentrations given are those measured prior to exposure to trypsin.

(A) HUVEC were grown with selected concentrations of trypsin-treated or -untreated pellet fractions (0 to 10 μg bacterial protein/ml). Proliferation was assessed by measurement of acid phosphatase activity after 5 days ($n = 12$, mean ± SD). The data are expressed as percent above control (no bacterial protein).

(B) Bacterial pellet fractions (50 μg protein/ml) were prepared with and without trypsin as described for panel A and placed in the lower compartment of the modified Boyden chamber. HUVEC migration was evaluated as described in Fig. 6 (PELLET fractions: $n = 2$, mean ± range; CONTROL without bacterial protein; $n = 4$, meand ± SD). (Adapted from Conely *et al.*[16])

to withstand oxidant stress likely contributes to survival within the host, enabling pathogenesis. The profound anemia associated with *B. bacilliformis* bacteremia is caused by adherence to, and subsequent invasion and lysis of erythrocytes. Lysis of erythrocytes leads to the local release and breakdown of hemoglobin. During this event, *B. bacilliformis* likely encounters local fluxes in $\cdot O_2^-$ generated by oxy-hemoglobin autoxidation associated with hemoglobin degradation following erythrocyte lysis.

Both *B. henselae* and *B. quintana* can cause persistent bacteremia. Although no association with circulating human erythrocytes or phagocytes has been proven definitively for *B. quintana* or *B. henselae*, recovery of such pathogens from blood is enhanced by the use of a blood culture system that lyses blood cells, indirectly suggesting such a state.[13,15,21,22] *In vivo, B. bacilliformis, B. henselae,* and *B. quintana* produce tissue lesions marked by infiltrates of inflammatory cells, including poly-morphonuclear and mononuclear phagocytes, in which the bacteria appear capa-ble of surviving and multiplying.[4,6,23,24] Although not yet directly measured, this would expose bacteria to significant amounts of oxygen-centered radicals result-ing from the respiratory burst of these activated phagocytes. Finally, tissue neo-vascularization, demonstrable in verruga peruana and bacillary angiomatosis, may increase local oxygen radical production as a by-product of locally increased oxygen tension.

In gram-negative bacteria, exposure to oxidant stress can induce a global response, which results in increased resistance to oxyradical-mediated killing.[25,26] A variety of antioxidant enzymes, DNA repair systems, and heat-shock proteins, that are under the positive control of distinct gene products, are induced as part of this global response (reviewed by Farr and Kogomo[27]). Direct scavenging of superoxide, hydrogen peroxide, and organic peroxides serves as a first-line defense against the toxic effects of oxygen-centered radicals. The metalloenzyme superox-ide dismutase (SOD) catalyzes the dismutation of superoxide radicals, minimizing the contribution of superoxide to oxyradical-mediated injury.[28] The isoenzyme members of the family include FeSOD, MnSOD, and CuZnSOD. Bacteria, with some exceptions, possess the MnSOD and/or FeSOD isoenzymes.[20] FeSOD appears to be constitutively expressed in most aerobic bacteria whereas MnSOD can also be induced under conditions of oxidative stress.[29,30] Isoenzymes are distinguishable on the basis of characteristic inhibition profiles (CuZnSOD is H_2O_2 and cyanide sensitive, FeSOD is H_2O_2 sensitive but not cyanide sensitive, and MnSOD is neither H_2O_2 nor cyanide sensitive), metal content, and amino acid and/or DNA sequence homology.[20] Several enzymes capable of hydrogen perox-ide and organic peroxide catabolism have been characterized. The selenoenzyme glutathione (GSH) peroxidase catalyzes the reduction of hydrogen peroxide or organic peroxides to their corresponding alcohols.[31] This activity is dependent on the availability of reduced glutathione maintained by the NADPH-dependent flavoprotein glutathione reductase. Although many purple bacteria possess sele-noenzymes as well as measurable amounts of GSH and GSH reductase activity, GSH peroxidase has not been described in these bacteria.[32] Catalase is a heme-containing peroxidase that converts hydrogen peroxide to oxygen and water, but cannot reduce organic peroxides. Although catalase activity has been described in many aerobic bacteria, it is often absent in organisms with limited capacity for *de novo* heme biosynthesis. It is noteworthy that *Bartonella* spp. exhibit heme-dependent growth *in vitro*,[13,21,33] suggesting limited capacity for *de novo* heme biosynthesis. A two-component alkyl hydroperoxide reductase (Ahp) enzyme system that catalyzes the conversion of organic peroxides to corresponding alcohols is described in *Escherichia coli* and *Salmonella typhimurium* (reviewed by Farr and Kogomo[27]), protect-ing these gram-negative bacilli from peroxide-induced mutagenesis by reducing a number of lipid peroxides. Although not directly involved in scavenging, glucose-6-phosphate dehydrogenase (G-6-PD), the activity of which is coupled with the reduction of NAD and NADP in prokaryotes, provides the necessary pool of reduced coenzyme for reduction of organic peroxides by either glutathione perox-idase or alkylhydroperoxide reductase.

Antioxidants clearly contribute to adaptation to and survival within an oxi-dative environment. In addition to playing a "housekeeping" role for the detox-ification of oxygen radicals generated during respiration, antioxidant enzymes do contribute directly to virulence in some pathogens. The presence of SOD activity in *Listeria monocytogenes* and *Nocardia* spp. has been demonstrated to be important in withstanding killing by the respiratory burst products of phagocytes.[34,35] Patho-genesis of the enteric gram-negative rod *Shigella flexneri* appears critically linked to the presence of SOD activity; SOD-deficient mutants are both more sensitive to phagocyte killing and less pathogenic.[36] The presence of antioxidants in intra-cellular pathogens is also thought to contribute to virulence. For example, the presence of antioxidant defenses in *Plasmodium* spp. is thought to play an impor-tant role for the survival of this protozoan in the oxygen-rich environment of the erythrocyte. In fact, several *Plasmodium* species have evolved the unique capacity to absorb host erythrocyte CuZnSOD, presumably augmenting superoxide scav-enging.[37]

3.2. Superoxide Dismutase Production by *B. henselae* and Other *Bartonella* spp.

Wack and collaborators[38,39] have reported superoxide radical scavenging by *Bartonella* spp. As measured both spectrophotometrically[40] and by activity bands after electrophoresis in native (nondenaturing) polyacrylamide gels,[41] SOD activ-ity was detected in whole-cell lysates of isolates of all *Bartonella* spp. that are proven human pathogens (*B. bacilliformis*, *B. quintana*, *B. henselae*, and *B. elizabethae*) as well as in *B. vinsonii* (generally considered an animal pathogen, but there has been one human blood isolate). Preliminary quantification of SOD by the spectrophotometric assay demonstrated large but variable amounts of activity (500–2000 U/mg) among the isolates tested. All isolates were found to possess a MnSOD-like (peroxide and cyanide insensitive) SOD activity (Fig. 9) that migrated as two or three closely grouped electromorphs at a low relative mobility (M_r) on native PAGE gels, dissimilar to purified *E. coli* MnSOD, which migrated as a single activity band at a relatively high M_r (Fig. 9). The M_r of SOD on native polyacrylamide gel varied by species (Fig. 9), suggesting natural differences in either native holoenzyme mass and/or charge among species. (A useful incidental application of such studies has been developed. When over 40 pathogenic clinical isolates of *B. henselae* and *B. quintana* were tested in a blinded fashion, all were speciated accurately based solely on their consistent species-specific SOD activity migration patterns on native polyacrylamide gels.)

The SOD activity of *B. henselae* has since been purified to greater than 95% homogeneity (data not shown). Analysis of the first 20 N-terminal amino acids confirmed it to be of the Fe/Mn SOD subclass, with sequence homology to the MnSOD of several *Halobacter* spp. The purified enzyme remained insensitive to hydrogen peroxide and sodium cyanide, consistent with previously described MnSODs. It was determined to have a subunit molecular mass of approximately 26,200 Da, and relatively high specific activity of 4400 U/mg. (Of note, the MnSOD isoenzyme is the inducible form of this enzyme class in both prokaryotes and eukaryotic mitochondria. Whether expression of MnSOD in *Bartonella* spp. is responsive to oxidant stress is under study.)

Little is known about the capacity of *Bartonella* spp. to scavenge hydrogen peroxide. Both *B. henselae* and *B. quintana* lack detectable catalase activity after growth *in vitro* (data now shown), as measured by techniques that stain for catalase activity on native polyacrylamide gels[42] or assay activity with an oxygen (Clark type) electrode.[43]

3.3. *B. henselae* Glucose-6-phosphate Dehydrogenase (G-6-PD) Activity

B. henselae does possess G-6-PD activity in cytoplasm-enriched fractions (Fig. 10). G-6-PD catalyzes the dehydrogenation of glucose-6-phosphate to 6-phosphogluconolactone, which is accompanied by the reduction of NADP. The availability of NADPH is essential for the catalytic reduction of hydrogen peroxide or organic peroxides via glutathione-dependent or Ahp-dependent reactions. In *E. coli* G-6-PD is inducible in response to oxidant stress, such as *in vitro* growth in the presence of a superoxide-generating agent like methyl viologen (paraquat).[44,45] Using a modification of a method for G-6-PD activity staining on native PAGE

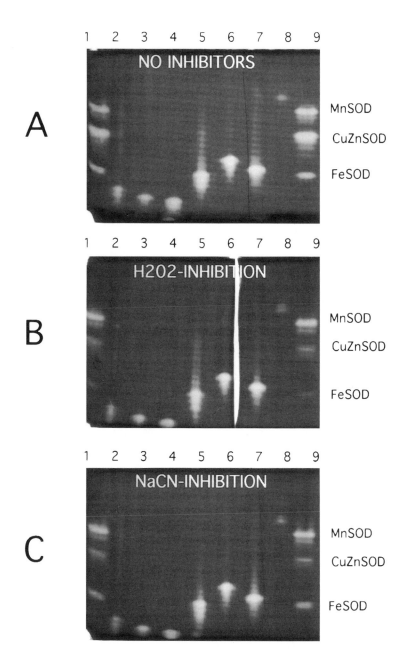

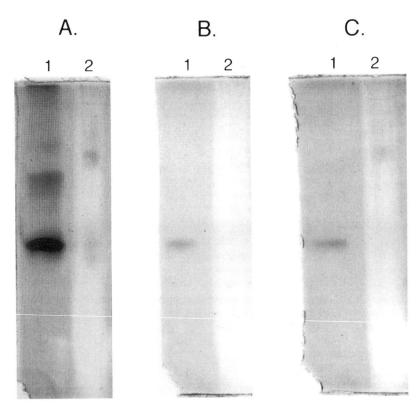

FIGURE 10. *B. henselae* glucose-6-phosphate dehydrogenase activity. Holoenzymes were separated on an 8% native lkpolyacrylamide gel prior to staining for G-6-PD activity. Land 1 in all gels contain 0.05 μg purified yeast G-6-PD, and lane 2 contain *B. henselae* whole lysate (65 μg/lane). Staining for G-6-PD activity was done in the presence of NADP and G-6-P (gel A), NADP without G-6-P (gel B), and NAD and G-6-P (gel C). Small amounts of G-6-PD activity are visualized in the lanes containing *B. henselae* whole lysate, but only in the presence of the necessary substrate, G-6-P. (Adapted from Wack *et al.*[38,39])

←

FIGURE 9. *Bartonella* spp. superoxide dismutase activities. SOD activity from a variety of representative isolates and purified enzyme (*E. coli* FeSOD, *E. coli* MnSOD, and bovine CuZnSOD) were visualized following gradient (5–15%) native PAGE, in the presence and absence of known inhibitors of FeSOD (H_2O_2) and CuZnSOD (H_2O_2 and cyanide). The SOD activity from whole lysates (15–30 μg/lane) of *B. bacilliformis* (ATCC No. 35685, lane 2), an animal isolate of *B. vinsonii* (ATCC No. VR 152, lane 3), a human blood isolate of *B. elizabethae* (ATCC No. 49927, referecned as the type strain, lane 4), a human blood isolate of *B. henselae* (lane 5), a human blood isolate of *B. quintana* (lane 6), the only known human blood isolate of *B. vinsonii* (lane 7), *Afipia felis* (ATCC No. 53690, lane 8), and purified *E. coli* FeSOD, *E. coli* MnSOD, and bovine CuZnSOD (lanes 1 and 9) are visualized after separation by native PAGE (gels A, B, and C). SOD activity in the absence of inhibitors is observed in all isolated studied, migrating as several bands at varying M_r in species-specific patterns (gel A). In duplicate gels (B and C), the major bands of SOD activity observed in all isolates are *not* inhibited by preincubation of the gels in either 1 mM H_2O_2 (gel B) or 2 mM Na cyanide (gel C), suggesting they are manganese-containing enzymes. As predicted, the *E. coli* FeSOD and the bovine CuZnSOD activities were H_2O_2 sensitive, and the CuZnSOD cyanide sensitive, whereas the MnSOD standard was cyanide and H_2O_2 insensitive under the same conditions. (Adapted from Wack *et al.*[33,39])

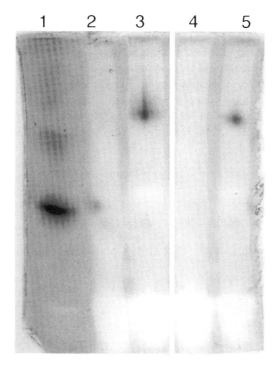

FIGURE 11. Induction of *B. henselae* G-6-PD activity: comparison of activity from *B. henselae* isolates grown in the presence and absence of paraquat. Two different *B. henselae* isolates were grown on chocolate agar in either the presence or absence of paraquat (10 and 30 μM, respectively). Whole lysates (40 μg) from each condition were separated on 8% native polyacrylamide gels, prior to staining for G-6-PD activity. Lane 1 is purified yeast G-6-PD, 0.05 μg. Lanes 2 and 3 are lysates (40 μg each) from the same isolate grown in the absence (lane 2), or presence of 10 μM paraquat (lane 3). Lanes 4 and 5 are lysates (40 μg each) from a second isolate of *B. henselae* grown in the absence (lane 4), or presence (lane 5) of 30 μM paraquat. A clear increase in G-6-PD activity is visualized in lanes containing *B. henselae* isolates grown in the presence of paraquat, suggesting induction of this enzyme in response to growth in this potentially oxidative environment. (Adapted from Wack *et al.*[38,39])

gels,[46] G-6-PD activity in *Bartonella* lysates has been visualized at a high M_r (Fig. 10), relative to purified yeast G-6-PD. No activity was visualized in the absence of either G-6-P or NAD, the characteristic substrate and coenzyme of G-6-PD, respectively. *Bartonella* G-6-PD also appeared to be either NADP or NAD dependent (data not shown) similar to this enzyme in other purple bacteria.

   B. henselae G-6-PD appears to be induced by oxidant stress. Activity was assayed in two *B. henselae* isolates after serial passage in the presence or absence of paraquat. As compared to the controls, a clear-cut increase in the G-6-PD activity demonstrable by native PAGE occurred in both isolates exposed to paraquat (Fig. 11). These preliminary enzyme activity data strongly suggest that G-6-PD activity represents an inducible antioxidant response in *B. henselae*. Quantitative comparisons of the changes in G-6-PD activity as well as analysis of this oxidant response at the transcriptional level are in progress.

In summary, *Bartonella* spp. do have significant capacity to catabolize superoxide radical, via catalytic dismutation by MnSOD, and possess an oxidant-inducible G-6-PD activity. Specifically how these organisms avoid hydrogen peroxide-induced toxicity, what other antioxidant mechanisms may be present, and whether antioxidants contribute to pathogenesis in the host, are areas of investigation currently under study.

REFERENCES

1. Allesandri, G., Raju, K., and Gullino, P., 1983, Mobilization of capillary endothelium *in vitro* by effectors of angiogenesis *in vitro*, *Cancer Res.* **43**:1790–1797.
2. Klagsbrun, M., and D'Amore, P. A., 1991, Regulators of angiogenesis, *Annu. Rev. Physiol.* **53**:217–239.
3. Folkman, J., and Shing, Y., 1992, Angiogenesis, *J. Biol. Chem.* **267**:10931–10934.
4. Arias-Stella, J., Lieberman, P. H., Erlandson, R. A., and Arias-Stella, J., Jr., 1986, Histology, immunohistochemistry, and ultrastructure of the verruga in Carrion's disease, *Am. J. Surg. Pathol.* **10**:595–610.
5. LeBoit, P. E., Berger, T. G., Egbert, B. M., Beckstead, J. H., Yen, Y. S. B., and Stoler, M. H., 1989, Bacillary angiomatosis. The histology and differential diagnosis of a pseudoneoplastic infection in patients with human immunodeficiency virus disease, *Am. J. Surg. Pathol.* **13**:909–920.
6. Cockerell, C. J., and LeBoit, P. E., 1990, Bacillary angiomatosis: A newly characterized, pseudoneoplastic, infectious, cutaneous vascular disorder, *J. Am. Acad. Dermatol.* **22**:501–512.
7. Kostianovsky, M., and Greco, M. A., 1994, Angiogenic process in bacillary angiomatosis, *Ultrastruct. Pathol.* **18**:349–355.
8. Garcia, F. U., Wojta, J., Broadley, K. N., Davidson, J. M., and Hoover, R. L., 1990, *Bartonella bacilliformis* stimulates endothelial cells *in vitro* and is angiogenic *in vivo*, *Am. J. Pathol.* **136**:1125–1135.
9. Garcia, F. U., Wojta, J., and Hoover, R. L., 1992, Interactions between live *Bartonella bacilliformis* and enothelial cells, *J. Infect. Dis.* **165**:1138–1141.
10. Relman, D. A., Loutit, J. S., Schmidt, T. M., Falkow, S., and Tompkins, L. S., 1990, The agent of bacillary angiomatosis: An approach to the identification of uncultured pathogens, *N. Engl. J. Med.* **323**:1573–1580.
11. Perkocha, L. A., Geaghan, S. M., Yen, T. S. B., Nishimura, S. L., Chan, S. P., Garcia-Kennedy, R., Honda, G., Stoloff, A. C., Klein, H. Z., Goldman, R. L., Meter, S. V., Ferrell, L. D., and LeBoit, P. E., 1990, Clinical and pathological features of bacillary peliosis hepatitis in association with human immunodeficiency virus infection, *N. Engl. J. Med.* **323**:1581–1586.
12. Relman, D. A., Falkow, S., LeBoit, P. E., Perkocha, L. A., Min, K.-W., Welch, D. F., and Slater, L. N., 1991, The organism causing bacillary angiomatosis, peliosis hepatitis, and fever and bacteremia in immunocompromised patients [letter], *N. Engl. J. Med.* **324**:1514.
13. Welch, D. F., Pickett, D. A., Slater, L. N., Steigerwalt, A. G., and Brenner, D. J., 1992, *Rochalimaea henselae* sp. nov., a cause of septicemia, bacillary angiomatosis, and parenchymal bacillary peliosis, *J. Clin. Microbiol.* **30**:275–280.
14. Slater, L. N., Welch, D. F., and Min, K.-W., 1992, *Rochalimaea henselae* causes bacillary angiomatosis and peliosis hepatis, *Arch. Intern. Med.* **152**:602–606.

15. Koehler, J. E., Quinn, F. D., Berger, T. G., LeBoit, P. E., and Tappero, J. W., 1992, Isolation of *Rochalimaea* species from cutaneous and osseous lesions of bacillary angiomatosis, *N. Engl. J. Med.* **327:**1625–1632.

16. Conley, T., Slater, L., and Hamilton, K., 1994, *Rochalimaea* spp. stimulate endothelial cell proliferation and migration *in vitro*, *J. Lab. Clin. Med.* **124:**521–528.

17. Halliwell, B., and Gutteridge, J., 1989, Oxygen is poisonous—An introduction to oxygen toxicity and free radicals, in: *Free Radicals in Biology and Medicine* (B. Halliwell and J. Gutteridge, eds.), Clarendon Press, Oxford, pp. 1–20.

18. Halliwell, B., and Gutteridge, J., 1989, Chemistry of oxygen radicals and other oxygen-derived species, in: *Free Radicals in Biology and Medicine* (B. Halliwell and J. Gutteridge, eds.), Clarendon Press, Oxford, pp. 22–81.

19. Misra, H. P., and Fridovich, I., 1972, The generation of superoxide radical during the autoxidation of hemoglobin, *J. Biol. Chem.* **247:**6960–6962.

20. Bannister, J., Bannister, W., and Rotillo, G., 1987, Aspects of the structure, function and applications of superoxide dismutase, *CRC Crit. Rev. Biochem.* **22:**111–180.

21. Slater, L. N., Welch, D. F., Hensel, D., and Coody, D. W., 1990, A newly recognized fastidious gram-negative pathogen as a cause of fever and bacteremia, *N. Engl. J. Med.* **323:**1587–1593.

22. Lucey, D., Dolan, M. J., Moss, C. W., Garcia, M., Hollis, D. G., Wegner, S., Morgan, G., Almeida, R., Leong, D., Greisen, K. S., Welch, D. F., and Slater, L. N., 1992, Relapsing illness due to *Rochalimaea henselae* in normal hosts: Implication for therapy and new epidemiologic associations, *Clin. Infect. Dis.* **14:**683–688.

23. Slater, L. N., Pitha, J. V., Herrera, L., Hughson, M. D., Min, K.-W., and Reed, J. A., 1994, *Rochalimaea henselae* infection in AIDS causing inflammatory disease without angiomatosis or peliosis: Demonstration by immunocytochemistry and corroboration by DNA amplification, *Arch. Pathol. Lab. Med.* **118:**33–38.

24. Min, K.-W., Reed, J. A., Welch, D. F., and Slater, L. N., 1994, Morphologically variable bacilli of cat scratch disease are identified by immunocytochemical labeling with antibodies to *Rochalimaea henselae*, *Am. J. Clin. Pathol.* **101:**607–610.

25. Demple, B., and Halbrook, J., 1983, Inducible repair of oxidative DNA damage in *Escherichia coli*, *Nature* **304:**466–468.

26. Christman, M., Morgan, R., Jacobson, F., and Ames, B., 1985, Positive control of a regulon for defenses against oxidative stress and some heat-shock proteins in *Salmonella typhimurium*, *Cell* **259:**5932–5936.

27. Farr, S. B., and Kogomo, T., 1991, Oxidative stress response in *Escherichia coli* and *Salmonella typhimurium*, *Microbiol. Rev.* **55:**561–585.

28. McCord, J. M., Keele, R. B., and Fridovich, I., 1971, An enzyme based theory of obligate anaerobiosis, *Proc. Natl. Acad. Sci. USA* **68:**1024–1027.

29. Hassan, H. M., and Moody, C. S., 1984, Induction of the manganese-containing superoxide dismutase of *Escherichia coli* by naladixic acid and iron chelators, *FEMS Microbiol. Lett.* **25:**233–236.

30. Hassan, H. M., and Fridovich, I., 1977, Regulation of the synthesis of superoxide dismutase in *Escherichia coli:* Induction by methyl viologen, *J. Biol. Chem.* **252:**7667–7672.

31. Flohé, L., 1982, Glutathione peroxidase brought into focus, *Free Radical Biol. Med.* **5:**223–253.

32. Fahey, R. C., and Sundquist, A. R., 1991, Evolution of glutathione metabolism, *Adv. Enzymol. Relat. Areas Mol. Biol.* **64:**1–53.

33. Schwartzman, W. A., Nesbit, C. A., and Baron, E. J., 1993, Development and evaluation of a blood-free medium for determining growth curves and optimizing growth of *Rochalimaea henselae*, *J. Clin. Microbiol.* **31:**1882–1885.

34. Welch, D., Sword, C., Brehms, S., and Dusanic, D., 1979, Relationship between superoxide and pathogenic mechanisms of *Listeria monocytogenes*, *Infect. Immun.* **23:**863–872.

35. Filice, G., 1983, Resistance of *Nocardia asteroides* to oxygen dependent killing by neutrophils, *J. Infect. Dis.* **148:**861–867.

36. Franzon, V., Arondel, J., and Sansonetti, P., 1990, Contribution of superoxide dismutase and catalase activities to *Shigella flexneri* pathogenesis, *Infect. Immun.* **58:**529–535.

37. Fairfield, A. S., Meshnick, S. R., and Eaton, J. W., 1983, Malaria parasites adopt host cell superoxide dismutase, *Science* **221:**764–766.

38. Wack, M. F., and Slater, L. N., 1994, Superoxide dismutase (SOD) activity in *Bartonella* (formerly *Rochalimaea*) *henselae* and *quintana*, *Clin. Infect. Dis.* **19:**571 (abstract #544, Annual Meeting of the Infectious Diseases Society of America).

39. Wack, M. F., Van de Wiele, C. J., Robinson, A. M., and Slater, L. N., 1996, Identification and purification of a manganese superoxide dismutase from *Bartonella*, submitted for publication.

40. Spitz, D., and Oberley, L., 1989, An assay for superoxide dismutase activity in mammalian tissue homogenates, *Anal. Biochem.* **179:**8–18.

41. Beauchamp, C., and Fridovich, I., 1971, Superoxide dismutase: Improved assay and an assay applicable to acrylamide gels, *Anal. Biochem.* **44:**276–287.

42. Sun, Y., Oberley, L., and Li, Y., 1988, A simple method for clinical assay of superoxide dismutase, *Clin. Chem.* **34:**497–500.

43. Metcalf, J., Gallin, J., Nauseef, W., and Root, R., 1986, Oxygen consumption, in: *Laboratory Manual of Neutrophil Function*, Raven Press, New York, pp. 98–100.

44. Kao, S. M., and Hassan, H. M., 1985, Biochemical characterization of a paraquat-tolerant mutant of *Escherichia coli*, *J. Biol. Chem.* **260:**10478–10481.

45. Greenberg, J. T., and Demple, B., 1989, A global response induced in *Escherichia coli* by redox-cycling agents overlaps with that induced by peroxide stress, *J. Bacteriol.* **171:**3933–3939.

46. Deutch, J., 1983, Glucose-6-phosphate dehydrogenase, in: *Methods of Enzymatic Analysis* (H. Bergmeyer, ed.), *Verlag Chemie, Weinheim, pp. 190–197.*

Index